Medical Technology

Milestones in Discovery and Invention

Medical Technology

LISA YOUNT

Facts On File, Inc.

Medical Technology

Facts On File, Inc.
11 Penn Plaza
New York, NY 10001

Library of Congress Cataloging-in-Publication Data

Yount, Lisa.
Medical technology / Lisa Yount.
p. cm. — (Milestones in discovery and invention)
Includes bibliographical references and index.
Summary: Identifies specific scientists and their contributions to some of the major advances in technology that have transformed surgery and medicine in the past two hundred years.
ISBN 0-8160-3568-7
1. Medical technology—History—Juvenile literature. 2. Medical scientists—Biography—Juvenile literature. [1. Medical technology. 2. Scientists.] I. Title. II. Series.
R855.4.Y68 1998
610'.9—dc21 97-17679

Facts On File books are available at special discounts when purchased in bulk quantities for businesses, associations, institutions or sales promotions. Please call our Special Sales Department in New York at (212) 967-8800 or (800) 322-8755.

You can find Facts On File on the World Wide Web at http://www.factsonfile.com

Text design by Cathy Rincon
Cover design by Smart Graphics
Illustrations on pages 39, 42, 57, 61, 67, 78, 80, 82, 88, 101, 114, 119, 121, 124, 127 by Jeremy Eagle

Printed in the United States of America

MP FOF 10 9 8 7 6 5 4 3 2

This book is printed on acid-free paper.

In memory of

DOUGLAS GORDON

Sometimes technology is not enough.

Contents

Introduction

Today, people facing surgery have little to fear. They know they will feel no pain during the operation. If they undergo major surgery, an anesthetic gas or injection will make them unconscious. If the surgery is more minor, a local anesthetic will numb the part of their body being worked on.

The instruments, bandages, and other material used during the operation will be sterile, or completely free of microorganisms. If the operation is major, the surgeons and their assistants will wear sterile clothing as well. These measures will greatly reduce the likelihood of infection.

Machines in the operating room will keep track of the patient's breathing, blood pressure, and other vital signs. They may even temporarily take over the work of the patient's heart and lungs. If the patient loses blood during the operation, transfusions will make up the loss. Images made by X rays or other imaging tools will guide the surgeon's hand, giving a preview of the inside of the patient's body before a cut is ever made.

Surgery was very different at the start of the 19th century. Straps or a husky assistant held patients down as they screamed in pain. Surgeons had no idea what they would find until they cut into a patient's body. Worst of all, as a contemporary surgeon wrote,

> Cleanliness was out of place. . . . The surgeon operated in a . . . coat of black cloth . . . stiff with the blood and filth of

> years. The more sodden [soaked with fluid] it was the more forcibly did it bear evidence to the surgeon's prowess [skill].

Patients who survived their nightmarish operating room ordeal were taken back to a filthy hospital ward. Most soon died there from infection. Loss of blood killed others before infection had time to set in.

Some of the changes that made today's safe, painless surgery possible grew out of basic discoveries in science. For example, Louis Pasteur's study of microorganisms in the mid-19th century led to discovery of the cause of wound infection. But it was medical technology—inventions that put science to practical use—that turned scientific breakthroughs into tools that could help sick or injured people. Scottish surgeon Joseph Lister took Pasteur's ideas from the laboratory to the operating room, working out practical ways to kill microorganisms in wounds. Some advances in medical technology, such as anesthesia, were even made without any scientific understanding of why they worked.

This book will describe some of the major advances in technology that have transformed surgery and medicine in the past 200 years. It will look at the people who created them, the scientific discoveries that inspired them, and the way doctors and the public reacted to them.

It will also highlight some differences between the advances in medical technology that were developed in the 19th and early 20th centuries and those developed in the last 50 years or so. Once early advances such as anesthesia, X rays, and blood transfusions were accepted, they became very widely used. Almost any doctor could use them on almost any patient. By contrast, many "high-tech" tools and procedures of today, such as medical imaging, artificial organs, and organ transplants, are available only in large hospitals. They are very expensive. They often benefit only the sickest patients or those with the best insurance. As a result, they raise ethical issues that earlier technological advances did not. Even as medical technology continues to advance, these issues cause problems that are likely to grow as the 21st century begins.

Further Reading

The following are some general books about 20th-century or current medical technology:

Fradin, Dennis B. *Medicine: Yesterday, Today, and Tomorrow.* New York: Children's Press, 1989. For young adults. Describes modern advances in medicine, including medical technology.

Frontiers of Medicine: Foundations for the Future. New York: Torstar Books, 1986. Well-illustrated book covers recent advances in medicine, including those using computers and other technology.

How Things Work: Medicine. Alexandria, Va.: Time-Life Books, 1991. Excellent, well-illustrated book provides detailed descriptions of the technology of modern medical diagnosis (discovery of causes of disease) and surgery, including medical imaging, artificial organs, and transplants.

Morris, Beryl. *Biotechnology.* Cambridge, England: Cambridge University Press, 1995. For young adults. Describes recent advances in medical technology, focusing on use of artificial parts to replace body parts or organs.

NOTE

pp. ix–x "Cleanliness was . . ." Quoted in Sherwin Nuland, *Doctors: The Biography of Medicine* (New York: Random House/Vintage Books, 1989), p. 347.

Key to Icons in Boxed Features

- Key Concepts
- Parallels
- Solving Problems
- Social Impact
- Connections
- Other Inventors
- I Was There
- Trends
- Issues

Medical Technology

1

Surgery Without Pain

WILLIAM MORTON AND ANESTHESIA

Although he was not the first to use anesthesia, ambitious American dentist William Thomas Green Morton was the one who made surgeons aware that there was a way to free people from pain during operations. (Courtesy National Library of Medicine)

One contemporary surgeon wrote that in the early 19th century, "a patient preparing for an operation was like a condemned criminal preparing for execution." With nothing except perhaps a little whiskey to dull the agony, even minor surgery was torture. The only help a surgeon could offer his screaming patients was speed. One London surgeon, for instance, bragged that he could cut off, or amputate, a leg in just 32 seconds. But what terrible seconds they were!

Ancient doctors knew of a few plant substances, such as dark sap or "tears" from the opium poppy, that would dull pain. Most of these compounds, though, could kill the patient if the dose was too high. They were ineffective if it was too low. Because of this

I WAS THERE

A SURGICAL NIGHTMARE

A patient whose leg was amputated in the days before anesthesia left this account of his ordeal:

> The operation . . . involved cruel cutting through inflamed and . . . sensitive parts, and could not be dispatched by a few strokes of the knife. Of the agony it occasioned I will say nothing. Suffering as great as I underwent cannot be expressed in words. . . . During the operation, in spite of the pain, my senses were preternaturally acute [supernaturally sharp]. . . . I still recall with unwelcome vividness the spreading out of the instruments, . . . the first incision [cut], the fingering of the sawed bone, . . . the tying of the blood-vessels, the stitching of the skin, and the bloody dismembered limb lying on the floor.

unpredictability, surgeons of the 18th and early 19th centuries gave up trying to prevent pain during operations. "The avoidance of pain while operating is an idea for a fairy tale and we should not concern ourselves with it any more," French surgeon Alfred Velpeau wrote in 1839. "The cutting of the knife and pain are two aspects of surgery which cannot be separated."

Laughing Gas Parties and Ether Frolics

The first hints that ending the agony of surgery might not be "an idea from a fairy tale" came from what today would be called drug parties. They started around 1795, when a 17-year-old British chemistry student named Humphry Davy and some fellow scientists experimented with gases by breathing them and noting the results. These young men thought that the gases might be used to treat lung ailments, but their experiments were done as much for fun as for science.

One of the gases Davy tried was a combination of nitrogen and oxygen called nitrous oxide. After breathing this gas, Davy felt so happy that he began shaking with uncontrollable laughter. He therefore gave nitrous oxide the nickname of "laughing gas."

Davy noticed that after he had breathed laughing gas, he sometimes bumped or bruised himself without feeling it. A few breaths of nitrous oxide also made the pain of his sore gums go away. In 1800, he wrote, "As nitrous oxide . . . appears capable of destroying physical pain, it may probably be used with advantage during surgical operations."

Neither Davy nor any surgeons of his day followed up on his idea. Young people looking for amusement did, however. Laughing gas parties became popular in Britain and the United States. Instead of laughing gas, some partygoers used another substance called sulfuric ether. Ether was a clear liquid that evaporated quickly into a strong-smelling gas. Its effects were much like those of nitrous oxide.

KEY CONCEPTS

anesthesia the state of being insensitive to pain, or a substance that produces such a state

ether a gas with effects similar to those of nitrous oxide

local anesthetic a substance that produces insensitivity to pain in part of the body without loss of consciousness

nitrous oxide a gas (N_2O) that produces intoxication and insensitivity to pain

Doctors and medical students were the most likely to take part in laughing gas parties or ether frolics. As Davy had predicted, both gases had come to be used as medical treatments for certain breathing problems. That meant that doctors could get supplies of the gases more easily than ordinary people could.

One young doctor, Crawford Williamson Long, became fond of inhaling ether while he was still a student. When he returned to his rural home in Jefferson, Georgia, he taught this habit to his friends. One day one of those friends, James Venable, complained to Long about a swelling on the back of his neck. Venable said he would like the lump removed but was afraid of the pain the operation would cause. Like Humphry Davy, Long had noticed that he and his fellow partyers sometimes bumped themselves or fell down, yet felt no pain. He suggested that if Venable breathed ether beforehand, the cutting might not hurt.

Venable agreed to try Long's idea. On March 30, 1842, he breathed ether in Long's office. He became drowsy and felt nothing when Long cut the small tumor from his neck. Indeed,

Venable refused to believe that the operation had taken place until Long showed him the tumor.

This operation was the first one in which ether or any other gas was used to deaden pain. Long did not feel sure enough of his results to publicize them, however. For the time being, the possibility of painless surgery remained buried in his little Georgia town.

"A New Era in Tooth Pulling"

While medical students had laughing gas or ether parties for fun, certain showmen used the same gases to make money. Traveling by horse and wagon through the countryside, they charged an admission fee to people who wanted to inhale the substances or watch others doing so.

One such showman, Gardner Quincy Colton, brought this questionable entertainment to Hartford, Connecticut, late in 1844. Among the people who flocked to Union Hall to watch their fellow citizens breathe laughing gas and make fools of themselves was a young dentist named Horace Wells.

As several intoxicated men lurched around the stage, Wells noticed that one of them, Sam Cooley, hit his shin hard as he fell over a bench. When Cooley sat down near him, Wells pointed out a deep gash in Cooley's leg. Only then did Cooley notice that he had been hurt. He told Wells that he had felt no pain at all from the cut.

This immediately gave Wells an idea much like the one that had struck Crawford Long. People feared dentists because of the pain of having a tooth extracted. Wells thought that if patients breathed nitrous oxide, extractions might not hurt. He met Colton after the show and arranged to buy a supply of gas from him. The next morning, December 11, Wells inhaled some of the gas and had his assistant pull a healthy tooth from his mouth. "It did not hurt so much as the prick of a pin," he exclaimed afterward. "This is a new era in tooth pulling!"

Unlike Long, Horace Wells wanted to spread the news of his discovery. In January 1845, he traveled to Boston, then the fourth-largest city in the United States and a center of medical science. The first person he called on was William Thomas Green Morton, another dentist. Morton, born in Charlton, Massachusetts, on August 9, 1819, had once been a partner in Wells's dental practice. After hearing Wells's story, Morton introduced Wells to John Collins Warren, the senior surgeon at Massachusetts General Hospital. Warren agreed to let Wells demonstrate his gas before an audience of medical students.

On the day of his demonstration, Wells gave nitrous oxide to one of the students, who had volunteered to have a sore tooth pulled. The young man appeared to fall asleep. As soon as Wells yanked on his tooth, however, he let out a piercing scream. In his haste, Wells had probably given the student too little gas before setting to work. "Humbug! Humbug!" the medical students shouted, sure now that Wells's treatment was a fake. Amid laughter, catcalls, and insults, the humiliated Wells crept out of the room.

The Ambitious Dentist

Wells's failure shamed him so much that he gave up not only laughing gas but his entire dental practice. His former partner, however, was made of sterner stuff. In spite of Wells's fiasco, which he had witnessed, William Morton believed that painless dentistry, or even painless surgery, might be possible. An ambitious young man, he was quick to see both the benefit to humanity and the profit to himself that such a technique might bring. Furthermore, unlike Long or Wells, he had the determination to stick with his research when he failed and to make sure everyone knew about it when he succeeded.

Morton discussed his ideas with Charles T. Jackson, a physician, geologist, and chemist who sometimes lectured at Harvard. Jackson suggested that Morton try ether instead of laughing gas. Taking Jackson's advice, Morton began experimenting on worms, goldfish, insects, and even his family dog—not to mention himself. He found that after the animals breathed ether, he could sometimes operate on them without their showing any sign of pain. The effects, however, varied considerably.

Reluctantly, because he did not want Jackson to start having ideas of his own, Morton asked the chemist's advice again. Jackson told him to use only pure ether, not the commercial version of the substance. In commercial ether, pure ether was often mixed with other chemicals that changed its effects.

Morton got his first chance to try ether on a human patient on the evening of September 30, 1846, when a music teacher named Eben Frost came to his office with a terrible toothache. Frost had heard that some dentists used hypnotism to prevent pain during tooth extractions, and he asked Morton to try this technique on him. Morton, however, said he had something better. Frost agreed to breathe ether before his tooth was pulled. The treatment worked so well that, like James Venable, Frost had trouble believing that the operation was over until he was shown its results.

The groggy Frost was hardly on his feet again when Morton dragged him to a notary and had him sign a statement about the operation under oath. Morton then proceeded to the office of the *Boston Daily Journal*, told the newspaper editor his story, and showed him Frost's statement. The next day the *Journal* printed the following announcement:

> Last evening, as we were informed by a gentleman who witnessed the operation, an ulcerated [diseased] tooth was extracted from the mouth of an individual without giving the slightest pain. He was put into a kind of sleep, by inhaling a preparation, the effects of which lasted for about three quarters of a minute, just long enough to extract the tooth.

Morton had a reason for not saying what his "preparation" was. On the same day the newspaper story appeared, he applied for a patent on his technique. Ether was a widely known chemical, so Morton could win a patent only if he kept the identity of his painkiller a secret.

Historic Operations

Morton now took an even bolder step. He asked John Warren, the surgeon who had witnessed Horace Wells's spectacular failure, for permission to use his sleep-inducing substance on a patient during one of Warren's operations. Amazingly, Warren agreed.

On October 16, 1846, the day Morton had promised to present his gas, Warren's operating room at Massachusetts General Hospital was packed with doctors, medical students, and curious members of the public. Most hoped to see another show as entertaining as the one Horace Wells had unintentionally put on.

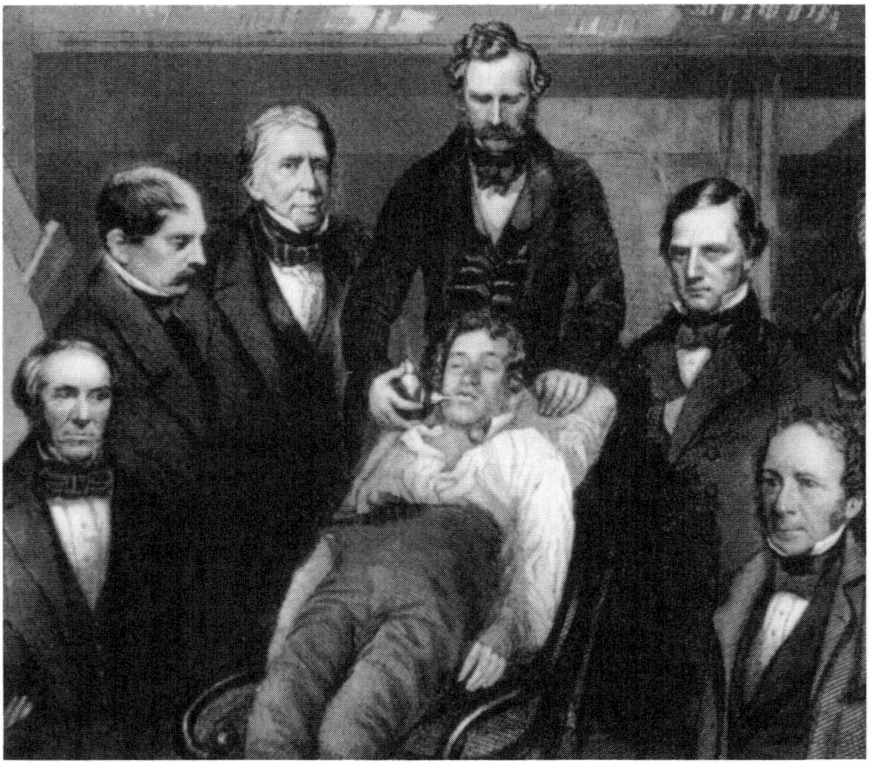

Morton put printer Gilbert Abbott to sleep before surgery at Massachusetts General Hospital by having Abbott breathe ether vapor from a glass globe. (Courtesy National Library of Medicine)

At first it seemed that the audience was doomed to disappointment: Morton did not appear. After waiting for 15 minutes, Warren prepared to start the operation without him. But just as the surgeon picked up his scalpel, Morton dashed through the door, carrying a glass globe that he planned to use in giving the gas. He explained that an instrument maker had finished the new device only a few minutes before. "Well, sir, your patient is ready," Warren said.

The patient was a young printer named Gilbert Abbott. He was going to have a large tumor removed from his lower jaw. Abbott began breathing from the globe, and after a few minutes he became unconscious.

As Warren made his first incision, the watching students braced themselves for a shriek like the one Wells's patient had given. It didn't come. Abbott did not move at all during the first part of the surgery. Toward the end of the 25-minute operation he became restless, but he said later that he felt only as if his neck had been scratched. When Warren put down his scalpel at last, the surgeon said solemnly to his audience, "Gentlemen, this is no humbug."

Word of Morton's innovation spread rapidly. Famed doctor and writer Oliver Wendell Holmes suggested calling it anesthesia, from a Greek word meaning "lack of feeling."

Henry Bigelow, another surgeon at Massachusetts General Hospital, wrote a report about Morton's demonstration that appeared in a Boston medical journal on November 18. Bigelow's father sent a copy of the article to a friend in England. The friend, in turn, told the story to Robert Liston, professor of surgery at University College in London.

Liston, like John Warren, was eager to try a possible method of surgical pain relief. On December 19, he carried out the first major operation performed under ether anesthesia in Europe. Before beginning the operation, an amputation, he told his audience of medical students, "We are going to try a Yankee dodge today, gentlemen, for making men insensible [unconscious]." At the operation's end, as impressed as Warren had been, Liston said, "This Yankee dodge, gentlemen, beats mesmerism [hypnotism] hollow."

A Bitter Wrangle

Back in the United States, the several discoverers and near-discoverers of anesthesia had begun a bitter wrangle over credit for the invention. That struggle would consume most of the rest of their lives.

On November 13, Charles Jackson, convinced that the original idea had been his, presented his claims in a letter to the Academy of Sciences in France. He thought that if his claim was accepted in Paris, then one of the centers of the medical world, he might become even more famous than if it was accepted in the United States. Shortly afterward, Horace Wells traveled to Paris with the same idea in mind.

OTHER INVENTORS

CHARLES JACKSON, CREATIVE CATALYST

Jackson's letter to Paris was not his first attempt to claim credit that he had not earned. Earlier he had unsuccessfully sued Samuel F. B. Morse, the inventor of the telegraph, demanding credit for that invention because of a talk that he and Morse had had. In that conversation Jackson had made a useful suggestion to Morse, just as he later did to Morton. But Morse and Morton, not Jackson, were the ones who turned these suggestions into working inventions.

In chemistry, a substance that helps a chemical reaction happen but does not take part in the reaction is called a catalyst. Jackson seems to have been a sort of inventors' catalyst. Unfortunately for him, when credit for inventions was handed out, the function of creative catalyst was not recognized.

OTHER INVENTORS

NO LONGER IN SORROW

James Young Simpson was a surgeon and obstetrician (doctor who treats women in childbirth) in Edinburgh, Scotland. He discovered that a gas called chloroform, like nitrous oxide and ether, could act as an anesthetic. In November 1847, he first used chloroform to ease the pains of a woman who was giving birth.

To Simpson's surprise, many religious leaders opposed his innovation. They objected to using painkillers during childbirth because the Bible's book of Genesis said that God had told Eve, "In sorrow shalt thou bring forth children." The outspoken Simpson replied that God Himself must have used anesthesia when he put Adam into what Genesis called a "deep sleep" before removing Adam's rib to create Eve.

The person who ended the opposition to childbirth anesthesia, though, was not Simpson but a woman: Britain's Queen Victoria. The beloved queen breathed chloroform while giving birth to her seventh child, Prince Leopold, on April 7, 1853. Her action transformed the use of childbirth anesthesia from a sinful act to the latest fashion.

The Academy of Sciences, France's chief scientific body, accepted Jackson's claim. The Paris Medical Society did the same for Wells. The French groups' opinions had little effect in the United States, however. There Morton remained supreme.

Morton's fame failed to bring him the riches he had hoped for, though. He won his patent, but the fact that he was merely using ether quickly leaked out, and the patent became worthless. Morton then changed his tack, sending petitions to Congress that asked for a large monetary reward for his "gift to humanity."

Members of Congress decided to award the discoverer of anesthesia $100,000, but they found it impossible to agree on who

PARALLELS

DISAPPOINTED INVENTORS

Anesthesia was not the only invention whose development was marked by frustrating struggles that eventually ruined the lives of most of the people involved. Radio, too, had this kind of unhappy history. Just as Jackson, Wells, and Morton competed for credit for the development of anesthesia, so Lee de Forest, Edwin Armstrong, and David Sarnoff fought over patents for inventions related to radio.
De Forest, like Morton, ended up a business failure. Armstrong, like Wells, committed suicide.

that discoverer was. Jackson entered his claims; so did the widow of Horace Wells. Even Crawford Long was eventually persuaded to get into the act. Finally, in 1849, Congress abandoned the whole issue. It was just too confusing.

Anesthesia may have ended pain for surgical patients, but it brought nothing but pain to most of its pioneers. Horace Wells's mind was destroyed by repeated experiments with anesthetic gases, and he killed himself in 1848. William Morton, bitter and worn out from his struggles to gain fame and money, died of a stroke in 1868, when he was only 49. Five years later, Morton's chief rival, Charles Jackson, was found raving beside Morton's grave. Jackson spent the remaining seven years of his life in an insane asylum. Crawford Long remained in Georgia and saw most of his way of life destroyed by the Civil War.

Only after their deaths were the contributions of these men fully recognized. Georgia chose Crawford Long to be one of its two citizens honored by statues in the Capitol Building in Washington, D.C. The people of Boston placed this inscription on William Morton's gravestone:

William T. G. Morton.
Inventor and Revealer of Anesthetic Inhalation.
By whom pain in surqery was averted and annulled.
Before whom in all time surgery was agony.
Since whom science has control of pain.

The Bostonians were probably right to give the honor to Morton. He was not the first to use anesthesia, but it was his ambition and perseverance that gave this invention to the medical world.

New Ways to End Pain

Within about a year of William Morton's groundbreaking demonstration at Massachusetts General Hospital, anesthesia came to be used in nearly all surgery performed in the United States or Europe. The blessing it brought to patients was obvious.

CONNECTIONS

THE HYPODERMIC SYRINGE

Some anesthetics are still inhaled as gases, but others are given by injection. Modern anesthesia thus would have been impossible without the hypodermic syringe, the familiar "shot needle." In 1853, a Frenchman, Charles Gabriel Pravaz, invented the hypodermic by attaching a hollow needle to a glass tube with a plunger at the other end. The tube held the liquid to be injected. When the plunger was pushed into the tube, it forced the liquid out through the needle and into the body.

Surgeons took a little longer to realize that it brought them a benefit as well: time. Now that they did not need to operate quickly to limit their patients' agony, surgeons could be more careful. They could also try operations that had seemed too lengthy and complex before.

Ether, chloroform, and nitrous oxide continued to be used in surgery well into the 20th century. For the most part, though, they have now been replaced by general anesthetics that produce fewer side effects, such as the injected anesthetic sodium pentothal. A general anesthetic makes a person completely unconscious.

For minor operations, doctors usually use a local anesthetic. Local anesthetics numb just the part of the body being operated on, leaving the patient awake but free of pain. Many local anesthetics are related to cocaine, which comes from a South American plant called coca. An Austrian physician, Carl Koller, first used cocaine as a local anesthetic in 1884.

Today, people who are to undergo a major operation usually receive four types of drugs. A tranquilizer makes them drowsy, a narcotic controls pain, a general anesthetic makes them lose

SOCIAL IMPACT

ADDICTION

Most narcotics (painkillers) and anesthetics have a dark side. In some doses they produce intoxication, or feelings of happiness and excitement. People may take the chemicals repeatedly to experience these effects. If they do, the substances change their brain chemistry so that they become dependent on, or addicted to, the chemicals. Addiction to chloroform ruined the mind of Horace Wells. Similarly, addiction to cocaine, heroin, or other narcotics destroys the lives of many people today.

consciousness, and a muscle relaxant paralyzes their muscles. (A mechanical respirator keeps them breathing.) Complex machines help the anesthetist, a surgical specialist, give the correct amount of each drug and keep track of the drugs' effects on breathing, blood pressure, and other vital functions.

It is hard to imagine the terror of surgery before anesthesia existed. Still, one can hardly argue with the words of Austrian doctor Josef Weiger, who wrote in the mid-19th century after seeing an operation performed under ether,

> This is the . . . greatest discovery of our century. I say the greatest discovery for, even though slowly, we could fulfil our other wishes though steamships and the electric telegraph had never been invented, [but] what will be gained by the prevention of pain in surgical operations can only be understood by those who have had to watch operations performed without anesthetics.

Chronology of Anesthesia

1800	Humphry Davy suggests that nitrous oxide be used to control pain in surgical operations
March 30, 1842	Crawford Long performs first operation using ether anesthesia
December 11, 1844	Horace Wells has tooth extracted under nitrous oxide anesthesia
January 1845	Wells fails in attempt to demonstrate nitrous oxide anesthesia in Boston
September 30, 1846	William Morton extracts tooth using ether
October 16	Morton demonstrates effectiveness of ether anesthesia in surgical operation

December 19	Robert Liston carries out first operation using anesthesia in Europe
November 1847	James Young Simpson uses chloroform anesthesia on woman in childbirth
1884	Carl Koller first uses cocaine as local anesthetic

Further Reading

Cartwright, Frederick F. *The Development of Modern Surgery.* New York: Crowell, 1967. Includes chapters on surgery in the early 19th century and on anesthesia.

Curtis, Robert H. *Triumph over Pain.* New York: David McKay Co., 1972. For young adults. Describes local and injected as well as inhaled anesthesia.

Fülöp-Miller, René (translated by Eden and Cedar Paul). *Triumph over Pain.* Indianapolis: Bobbs-Merrill, 1938. Dramatic story of the discovery and early development of anesthesia. Some dialogue is fictionalized but documents are accurate.

Gales, Judith. *Anesthetics: Surgery Without Pain.* San Diego, Calif.: Lucent Books, 1992. For young adults. Describes the development of anesthesia and different kinds of anesthetics.

Glaser, Hugo. *The Road to Modern Surgery.* New York: Dutton, 1962. Chapter on the development of anesthesia includes description of anesthetics used in the mid-20th century.

Nuland, Sherwin B. *Doctors: The Biography of Medicine.* New York: Random House/Vintage Books, 1989. Includes interesting chapter on the development of anesthesia.

Virtual Museum of Anesthesiology. Illustrated overview of the history of anesthetics, including references. http://umdas.med.miami.edu/aha/vma.

NOTES

p. 2 "a patient . . ." Quoted in René Fülöp-Miller (translated by Eden and Cedar Paul), *Triumph over Pain* (Indianapolis: Bobbs-Merrill, 1938), p. 90.

p. 2 "The operation . . ." Quoted in Fülöp-Miller, p. 89.

p. 3 "The avoidance . . ." Quoted in Hugo Glaser, *The Road to Modern Surgery* (New York: Dutton, 1962), p. 12.

p. 3 "As nitrous oxide . . ." Quoted in Sherwin B. Nuland, *Doctors: The Biography of Medicine* (New York: Random House/Vintage Books, 1989), p. 274.

p. 5 "It did not hurt . . ." Quoted in Frederick F. Cartwright, *The Development of Modern Surgery* (New York: Crowell, 1967), p. 29.

p. 5 "This is a new era . . ." Quoted in Nuland, p. 282.

p. 6 "Humbug! Humbug!" Quoted in Nuland, p. 283.

p. 7 "Last evening . . ." Quoted in Nuland, p. 287.

p. 9 "Gentlemen, this is no humbug." Quoted in Nuland, p. 289.

p. 9 "We are going to try . . ." "This Yankee dodge . . ." Quoted in Nuland, p. 293.

p. 13 "William T. G. Morton . . ." Quoted in Nuland, p. 301.

p. 15 "This is . . ." Quoted in Fülöp-Miller, p. 168.

2

Clean Wounds

JOSEPH LISTER AND ANTISEPSIS

British surgeon Joseph Lister found a way to prevent wound infections by using chemicals to kill microorganisms that got into wounds. (Courtesy National Library of Medicine)

When London surgeon Robert Liston first tried the "Yankee dodge" of anesthesia in an operation in 1846, one member of his fascinated audience was probably a serious-looking young man named Joseph Lister. Lister wasn't a doctor, or even a medical student, yet; he was still completing his bachelor's degree at University College. From childhood, however, he had planned to be a surgeon, and he would not have wanted to miss an operation as exciting as this one promised to be.

Hospital Diseases

Like the other members of the crowd and Liston himself, Lister would have realized on that day that he was witnessing a miracle: the end of surgical pain. Yet he also surely knew that, in some ways, the miracle was only temporary. The agony that the man on the operating table was being spared now would probably visit him soon in another form. His amputation wound would turn red, ooze milky pus, and give off a sickly sweet smell: all sure signs of infection. Just such a wound infection had made this amputation necessary in the first place. If the man suffered the fate of almost half the surgical patients in most hospitals, the new infection would spread through his body, burn him up with fever, and kill him in a few days.

No one knew what caused wound infections. Surgeons accepted them as a fact of life, just as they had accepted pain before anesthesia was introduced. They did know, though, that wounds were more likely to become infected inside hospitals than outside. James Young Simpson, the surgeon who had introduced chloroform anesthesia, estimated that 41 percent of surgical patients in hospitals with more than 300 beds developed such infections. By contrast, only 11 percent of people who had the same operations in their homes or in doctors' offices in the country got them. Simpson and other surgeons even called the several types of

I WAS THERE

"WHO IS TO PAY FOR THE FUNERAL?"

Sir Frederick Treves, a contemporary British surgeon, wrote:

> The attitude that the public assumed toward hospitals and their works . . . may be illustrated by the following incident. I was instructed by my surgeon to obtain a woman's permission for an operation on her daughter. The operation was one of no great magnitude [not very serious]. I interviewed the mother in the Receiving Room. I discussed the procedure with her in great detail. . . . I [then] asked her if she would consent to the performance of the operation. She replied: "Oh! it is all very well to talk about consenting, but who is to pay for the funeral?"

wound infection "hospital diseases." "The man laid on the operating table in one of our surgical hospitals," Simpson wrote grimly, "is exposed to more chances of death than was the English soldier on the field of Waterloo [a famous battle]." It was no wonder that people saw admission to a hospital as a death sentence. Only the poorest, who could get medical care in no other way, went there.

A Successful Young Surgeon

Joseph Lister would possess the blend of scientific insight and quiet determination that was needed to change this picture. Born on April 5, 1827, to wealthy English wine merchant Joseph

Jackson Lister and his wife, Isabella, Lister was raised as a member of the Religious Society of Friends, commonly called the Quakers. This group believed that each person should live according to his or her "inner light" rather than being swayed by the views of others. When Lister found a way to prevent deaths from wound infection, his Quaker upbringing would give him the strength to keep pushing his ideas in spite of powerful opposition.

Science was as much a part of Joseph Lister's childhood as was religion. His father had found a way to make the compound microscope (a microscope with two lenses) a useful scientific tool for the first time since its invention around 1590. Young Joseph learned how to use his father's microscopes. He also learned about the bodies of animals by cutting up dead ones that he found. He decided to be a surgeon when he grew up.

As a young man, Lister studied medicine at University College in London. He took his medical degree with honors in 1852. His professors then advised him to visit several hospitals to gain experience. They particularly recommended the hospital in Edinburgh, the capital of Scotland.

CONNECTIONS

FROM LISTER TO PASTEUR TO LISTER

Joseph Jackson Lister worked out the best distance by which to separate the two lenses of a compound microscope. Use of his formula removed distortions of shape and color in the microscope image that had kept the microscope from showing the world of the very small as it really was. In the mid-19th century, French scientist Louis Pasteur used compound microscopes improved by Lister's technique to study microorganisms, or microscopic living things. Pasteur's discoveries, in turn, gave Joseph Lister the clue he needed to find the cause and prevention of wound infection.

Lister arrived in Edinburgh in September 1853, intending to stay only a few months. He began studying under James Syme, the professor of clinical surgery at the University of Edinburgh's medical school. The two men became friends, and Lister decided to stay in Edinburgh. He soon was working under Syme as a

OTHER INVENTORS

IGNAZ SEMMELWEISS AND CHILDBED FEVER

One person who had tried to stop infection before Lister's day was a Hungarian doctor named Ignaz Semmelweiss. In 1846, as a young man, Semmelweiss worked in the obstetrics wards of a large hospital in Vienna, Austria. There he saw many women die of raging fevers after giving birth to their babies. He noticed that this so-called childbed fever, common in all hospitals, occurred much more often in the ward staffed by medical students than in the ward staffed by midwives, women who were not doctors but had been trained to help in the birth process.

One day a friend of Semmelweiss's accidentally cut his hand in the room where medical students dissected dead bodies as part of their training. He died a few days later of an illness that strongly resembled childbed fever. Seeing this, Semmelweiss guessed that whatever caused the fever might be carried to women on the hands of the students, who often went straight from the dissecting room to the obstetrics ward. He ordered all students to wash their hands in water containing a chlorine compound, already in use as a disinfectant, before entering the ward. The death rate on the ward dropped sharply. Other doctors nonetheless refused to adopt Semmelweiss's methods, and his ideas were soon forgotten.

surgeon in the Royal Infirmary, Edinburgh's chief hospital. He married Syme's oldest daughter, Agnes, in April 1856.

In 1860, Lister became professor of surgery at the University of Glasgow, Scotland, and a surgeon at Glasgow's Royal Infirmary. There he found the death rate from wound infection even higher than it had been in Edinburgh. By this time he was convinced that, contrary to what others thought, infection was not a normal part of wound healing. As a first attempt to fight it, he ordered the doctors and nurses who worked under him to wash their hands before and after touching patients' open wounds. A few other surgeons in the past had done the same. Because these measures did not completely eliminate the cause of the infections, however, they had little effect.

"The Beautiful Researches of Pasteur"

Lister had no idea what caused wound infections until the end of 1864, when a fellow professor showed him some papers written by a French chemist named Louis Pasteur. Pasteur's work was just starting to make a stir in his own country. Few British scientists had heard of him. Pasteur had discovered that fermentation, the process that creates alcohol in wine and beer, was caused by microscopic living things—germs, as he called them. Other kinds of germs caused putrefaction, or decay of dead animal tissue.

Many kinds of germs, Pasteur said, lived on tiny specks of dust in the air. Fermentation and putrefaction could take place only when such dust came into contact with plant or animal matter. Pasteur showed that if meat broth, which normally spoiled quickly, was first boiled to kill any germs in it and then kept in flasks constructed so that air could get in but dust could not, it stayed unspoiled for years.

After reading what he later called "the beautiful researches of Pasteur," Lister realized that the decay of tissue in wound infections was much like the putrefaction of dead bodies or spoiling

The discovery of French chemist Louis Pasteur (shown here) that microorganisms in the air caused the decay of meat gave Joseph Lister the clue he needed to guess the cause of wound infections. (Courtesy National Library of Medicine)

KEY CONCEPTS

antisepsis prevention of infection by killing germs that get into wounds
asepsis prevention of infection by keeping germs out of wounds
carbolic acid a strong chemical made from coal tar that can kill germs
germ a microscopic living thing; microorganism; microbe

meat. If germs in the air caused putrefaction, he reasoned, they might well cause wound infections, too.

Keeping all dust out of wounds did not seem practical. Lister thought, however, that he might find a way to kill the germs that got into wounds. As he wrote later,

> Just as we may destroy lice [tiny parasites] on the head of a child . . . by poisonous applications which will not injure the scalp, so, I believe, we can use poisons on wounds to destroy bacteria without injuring the soft tissues of the patient.

A story Lister heard about the Scottish city of Carlisle led him to the substance he needed. At that time, Carlisle, like many other cities, dumped its sewage into the streams that watered nearby farmers' fields. That produced two annoying problems: a terrible smell, and sickness among the cattle that grazed in the fields. The sickness was caused by microscopic parasites that lived in the sewage. To stop these problems, someone suggested sprinkling the polluted fields with carbolic acid, or phenol, a strong-smelling chemical made from coal tar. The Carlisle city fathers tried this treatment in 1864, and it worked.

Lister guessed that the carbolic acid had killed the germs responsible for the smell and the cattle sickness. He knew that solutions of carbolic acid could be put on human skin without seriously harming it. He decided that this chemical might be the germ killer he was looking for.

The Antiseptic Principle

Lister decided to try soaking wound dressings, or cloth coverings, in carbolic acid. At first he used this experimental treatment on only one kind of injury: compound fractures. In a simple fracture, a bone is broken but the skin remains undamaged. Such fractures normally heal without infection. In a compound fracture, however, the bone pokes through the skin, creating an open wound. In Lister's time, compound fractures almost always became infected.

One of the first compound fracture patients Lister treated was James Greenlees, an ll-year-old boy whose leg had been run over by a cart. Greenlees was brought to the Royal Infirmary on August 12, 1865. Lister covered his wound with lint (shredded cloth) soaked in carbolic acid. The wound healed without any sign of infection.

Greenlees's wound had been fairly small, and Lister knew it might have healed cleanly without special treatment. That was not the case with two other children he treated a few months later, however. Their bones had been broken by machines in factories where they worked, as many poor children had to do at the time. One boy's forearm was so badly torn that strips of muscle hung down from it. Yet these wounds, too, healed after the carbolic acid treatment.

After he had treated 11 compound fracture cases, Lister felt sure enough of his results to write an article about them. Titled "On a New Method of Treating Compound Fractures," it appeared in the *Lancet*, Britain's top medical journal, in 1867. Lister reported that nine of his cases had healed without infection after the carbolic acid treatment. He called his treatment antisepsis, meaning "against infection," because its purpose was to kill microbes in wounds in order to prevent the infections they caused.

Lister's article caused little stir. Most of the doctors who read it thought it just described a new technique for dressing wounds. They did not recognize antisepsis as the fundamental advance it was.

Lister next applied his technique to other kinds of accidental wounds. The results were equally successful. Then, in April 1867, he began using it on wounds caused by surgery. The first operations to which he applied it were simple ones, the draining of abscesses. In these local infections, pus accumulates under the skin, causing a painful swelling. Such swellings have to be opened and drained before they can heal. One of Lister's abscess patients was Queen Victoria, a brave woman who again showed herself willing to try the newest surgical technology.

Later in 1867, when he felt he had treated enough cases to show without question that his new method worked, Lister described it in detail in a medical paper called *On the Antiseptic Principle of the Practice of Surgery*. In this paper he pointed out that before he had begun using carbolic acid, his ward at the Glasgow hospital had had more infections than anyone else's. However, he wrote:

> since the antiseptic treatment has been brought into full operation, and wounds and abscesses no longer poison the atmosphere with putrid [decaying] exhalations, my wards, though in other respects under precisely the same circumstances as before, have completely changed their character; so that during the last nine months not a single instance of pyaemia, hospital gangrene, or erysipelas [different types of wound infection] has occurred in them.

A Chilly Reception

Most of Lister's fellow surgeons paid no more attention to this paper than they had to his earlier article. Those who did comment usually attacked it. They pointed out that others before Lister had

PARALLELS

ANESTHESIA AND ANTISEPSIS

Anesthesia and antisepsis both were discovered independently several times, yet were ignored by the medical establishment until they were adopted by someone persistent and publicity-wise enough to make others listen: Morton in the case of anesthesia and Lister in the case of antisepsis. Both also had a less successful inventor whose life ended tragically: Ignaz Semmelweiss, like Horace Wells, could not survive the ridicule with which his ideas were received and eventually went insane. Ironically, Semmelweiss died of infection in a wound he accidentally gave himself with a scalpel.

recommended carbolic acid (a fact Lister never denied) and that it had not worked. In vain Lister replied that carbolic acid was not the point. Other chemicals, he said, might serve equally well or even better. The important thing to understand was the necessity of killing the germs that got into wounds.

Most of those who did understand Lister's point did not agree with it. If they had heard of Pasteur's "germ theory" of disease at all, they disbelieved it. Even if they granted that germs existed, most doctors could not understand how such tiny living things could cause illness. Many probably also could not face the thought that, if Pasteur and Lister were right, they themselves had unwittingly killed many of their patients by introducing germs into their wounds.

Many surgeons and nurses simply thought Lister's techniques were too much bother. Even those who were willing to try his methods usually did not follow his complex directions exactly. Infection therefore continued to occur, and patients continued to die. Lister's critics then had even more evidence for their claim that his new technique was a failure. Still, Lister refused to give up.

In Contact with Genius

In 1869, James Syme suffered a stroke and had to retire. The University of Edinburgh offered Syme's job to Lister, and he accepted. He remained in Edinburgh for 11 years, during which time both he and his theories became very popular in the city. One of his adoring students wrote, "We knew we were in contact with Genius. We felt we were helping in the making of history and that all things were becoming new." In the wider surgical world, however, arguments between "Listerians" and "non-Listerians" continued. Science historian Sherwin B. Nuland writes that during this period, Lister "became one of the best known and most controversial scientists in the world."

Lister's ideas gained converts on the European continent much more quickly than they did in Britain or America. Many German and, to a lesser extent, French surgeons began applying his methods soon after he first described them. When Lister and his wife traveled through the Continent in 1875 to see how different hospitals applied his method, *Lancet* reported that the trip became a "triumphal march."

During his years in Edinburgh, Lister steadily refined his treatment. For instance, he found that a strong solution of carbolic acid irritated patients' skin and could even be poisonous. He therefore added more water to the solution he used in bandages. He also blended carbolic acid into a paste that could be used to cover large wounds.

An operation by Lister or any other surgeon who followed his methods became an elaborate ritual. Before surgery began, the walls and floors of the operating room were washed with carbolic acid solution. Surgical instruments, dressings, and everything else used in the operation were soaked in a similar solution or treated with carbolic acid lotion. The surgeon washed his hands in carbolic lotion and spread it on the patient's skin.

As if all this were not enough, all through the surgery a large pump sprayed the operating room with carbolic acid mist to kill germs floating in the air. Even some of Lister's supporters balked

CONNECTIONS

BETTER THREAD FOR SURGEONS

To stop bleeding, surgeons must tie shut the blood vessels they cut through during an operation. They also must sew up wounds and surgical cuts, or incisions. In Lister's time they used silk thread or thin metal wires for these purposes. The ends of the thread or wire were left outside the skin when the incision was closed so that they could be removed later. These dangling ends provided an easy path by which germs could enter the body.

In 1868, Joseph Lister began using catgut instead of silk or wire for surgical thread. This material, made from the intestines of sheep and cattle, had until then been used mainly to make strings for musical instruments. Surgical catgut could be cut short and left inside the body, where it slowly dissolved. Sherwin Nuland calls catgut thread "one of the most practical innovations ever made in . . . surgery."

at being surrounded by such a foul-smelling cloud. Most surgeons stopped using the spray around 1880, when scientists found that germs in the air (as opposed to those on surfaces) were usually harmless. Lister himself, however, insisted on it for 10 more years.

A Missionary for Antisepsis

Lister left the University of Edinburgh in 1877 to become professor of clinical surgery at King's College, London. Most of his followers could not understand why he abandoned this famous school, where he was so highly respected, to go to a lesser university in London, where many of his fiercest critics lived. His biographers, however, think that was precisely why he went. Edinburgh had been converted to his ideas, but London had not. As a kind of missionary for antisepsis, he felt a duty to change that.

Lister's reception in London was colder than anything he had encountered before. His hospital nurses made it clear that they resented being told how to run their wards. Other surgeons made jokes about his ideas. For instance, one cried with mock horror whenever someone came into his operating room, "Shut the door quickly in case one of Mr. Lister's microbes comes in!"

The evidence supporting Lister's ideas became harder and harder to ignore, however. In 1876, a German doctor named Robert Koch proved beyond question that a particular microorganism caused a particular disease. Two years later, Koch published a paper in which he linked six kinds of wound infection to six different kinds of microbes. Convinced by these proofs and by

I WAS THERE

LITTLE ACCIDENTS

Some of Lister's staff members from Edinburgh came with him to London. Even when faced with people who opposed or did not understand antiseptic techniques, they found ways to put those techniques into practice. For instance, one wrote:

> Not long after we arrived one of the surgeons was going to amputate a limb, and he asked me to come and look after the antiseptic arrangements. His technique was pretty defective, and he did not seem able to visualize the germs which were ready to seize every opportunity of getting into the wound, so I accidentally from time to time squeezed a quantity of carbolic lotion over his hand, for which I was, of course, very apologetic.

the drop in death rate in the wards run by Lister and others who followed his techniques, English and American surgeons finally began to accept Lister's ideas in the 1880s and 1890s.

Ironically, by the time the more reluctant surgeons were getting around to taking up Lister's approach, other surgeons and scientists, including Louis Pasteur himself, were moving beyond it. In 1878, Pasteur told doctors at the Academy of Medicine in Paris to follow techniques not of antisepsis but of asepsis—keeping germs out of the body rather than killing them once they got in.

By the 1890s, asepsis was combining with or replacing antisepsis in many European and American operating rooms. Surgeons began using the aseptic techniques that they still use today, such as wearing caps, masks, gowns, and gloves that are thoroughly cleaned or discarded after each operation. Heat or other means were used to make surgical instruments and anything else that would touch the patient sterile, or completely free of microorganisms.

"Humanity Salutes You"

Joseph Lister died on February 10, 1912. He had received many honors in his later years, even being elevated to the nobility. More important to him, however, were the lives his techniques saved and the changes those techniques made in surgery.

Anesthesia had given surgeons time to perform complex operations. Antisepsis and asepsis gave them the freedom to bring healing into body cavities that they had never dared to enter before. In the mid-1860s, British surgeon Sir John Erichsen had said that the abdomen, the chest, and the skull would be "forever shut from the intrusion of the wise and humane surgeon" because opening them was sure to bring on the patient's death by infection. By the 1890s, however, surgeons were operating in all these areas. Along with anesthesia, Joseph Lister's antisepsis created modern surgery.

Although some of his contemporaries resisted Lister's ideas, others recognized his achievement. When he attended the International Medical Congress in Amsterdam in 1879, according to an eyewitness's report,

> Professor Lister was received by the whole Congress with an enthusiasm which knew no bounds. When he stepped forward . . . the whole assembly rose to their feet; and, with deafening and repeated rounds of cheers, waving of hats and handkerchiefs, hailed the distinguished Professor of King's College. . . . Professor Donders, the president [of the congress], . . . said: "Professor Lister, it is not only our admiration which we offer you; it is our gratitude, and that of the nations to which we belong."

Late in the 19th century, when even the United States had finally become converted to antisepsis, the American ambassador to England took Donders's sentence a step further. He said to Lister, "My lord, it is not a Profession, it is not a Nation, it is Humanity itself which, with uncovered head, salutes you."

Chronology of Lister and Antisepsis

April 5, 1827	Joseph Lister is born in Britain
1852	Obtains medical degree from University College, London
1854	Begins work at Royal Infirmary in Edinburgh
1860	Becomes professor of surgery at University of Glasgow
1865	First uses antiseptic (carbolic acid) treatment
1867	Publishes papers on antiseptic treatment
1869	Becomes professor of clinical surgery at Edinburgh University

1870s	Antiseptic methods widely accepted in Continental Europe
1877	Lister becomes professor of clinical surgery at King's College, London
1880s	Antiseptic methods widely accepted in England
1890s	Antiseptic methods accepted in United States Aseptic methods begin to replace antiseptic methods Surgeons operate in abdomen, chest, and skull
February 10, 1912	Lister dies

Further Reading

Clendening, Logan, ed. *Source Book of Medical History*. New York: Dover reprint, 1960. Includes writings by Lister and others describing antiseptic and aseptic techniques. Hard reading.

Farmer, Laurence. *Master Surgeon*. New York: Harper & Brothers, 1962. Short, readable biography of Lister.

McTavish, Douglas. *Joseph Lister*. New York: Bookwright Press, 1992. Short biography of Lister for young readers.

Nuland, Sherwin B. *Doctors: The Biography of Medicine*. New York: Random House/Vintage Books, 1989. Contains an excellent chapter on Lister.

Truax, Rhoda. *Joseph Lister, Father of Modern Surgery*. Indianapolis: Bobbs-Merrill, 1944. Older biography of Lister for young adults.

Thorwald, Jürgen. *The Century of the Surgeon*. New York: Bantam Books, 1963. Contains a section on Lister and antisepsis. Includes some fictionalized dialogue, but facts and quotations of written material are accurate.

NOTES

p. 20 "The attitude . . ." Quoted in Sherwin B. Nuland, *Doctors: The Biography of Medicine* (New York: Random House/Vintage Books, 1989), p. 348.

p. 20 "The man laid . . ." Quoted in Laurence Farmer, *Master Surgeon* (New York: Harper & Brothers, 1962), p. 42.

p. 23 "The beautiful . . ." Quoted in Farmer, p. 46.

p. 25 "Just as we may . . ." Quoted in Nuland, p. 364.

p. 27 "since the antiseptic treatment . . ." Quoted in Logan Clendening, ed., *Source Book of Medical History* (New York: Dover reprint, 1960), p. 621.

p. 29 "We knew we were . . ." Quoted in Sarah R. Riedman, *Masters of the Scalpel* (Chicago: Rand McNally, 1964), p. 160.

p. 29 "became one of the . . ." Nuland, p. 368.

p. 29 "triumphal march." Quoted in Farmer, p. 96.

p. 30 "one of the most . . ." Nuland, p. 382.

p. 31 "Shut the door . . ." Quoted in Farmer, p. 111.

p. 31 "Not long after . . ." Quoted in Riedman, pp. 169–70.

p. 32 "forever shut . . ." Quoted in Farmer, pp. 119–20.

p. 33 "Professor . . ." Quoted in Farmer, pp. 112–13.

p. 33 "My lord . . ." Quoted in Nuland, p. 349.

3

Skeleton Rays

WILHELM RÖNTGEN AND X RAYS

German physicist Wilhelm Röntgen discovered "a new kind of ray" that gave doctors a way to see inside the body without surgery. (Courtesy American Institute of Physics, Emilio Segrè Visual Archives, Lande Collection)

Professor Röntgen wouldn't come to dinner. The maid went downstairs to call him several times, but he refused to leave his laboratory. When he finally did emerge, he said nothing, ate a few bites, and disappeared again.

Röntgen's wife, Bertha, was used to her husband being preoccupied with his work, but she had never seen him like this. She thought he must be upset because something had gone wrong with one of his experiments. In fact, however, Wilhelm Röntgen was puzzled and excited rather than angry. On the day of the missed dinner—November 8, 1895—he had seen something in his laboratory that made meals, or just about anything else, seem completely unimportant.

A Mysterious Glow

Röntgen, a 50-year-old professor of physics at the University of Wurzburg in Germany, often spent long hours in his laboratory rooms in the university's Physical Science Institute. Starting about a year and a half before this momentous day, he had used his lab time for experiments on certain odd new phenomena connected with electricity.

A British scientist named William Crookes had created a glass tube containing two electrodes through which an electric current could be passed. Most of the air was pumped out of the tube, producing an almost complete vacuum (absence of gas) inside. When a high-voltage current was passed through a Crookes tube, as this invention was known, rays coming from the cathode, or negatively charged electrode, made the tube glow. Crookes had described these "cathode rays" in 1878. Since that time, Crookes and other physicists, including Röntgen, had been trying to learn more about them.

KEY CONCEPTS

cathode rays rays sent from the cathode, or negatively charged electrode, of a Crookes tube

Crookes tube sealed glass tube with vacuum inside, containing two electrodes of opposite charge

radiology the medical specialty of using X rays and other radiation to diagnose and treat disease

X rays form of high-energy electromagnetic radiation that penetrates most materials but is stopped by bone

Röntgen wanted to know whether the Crookes tube produced any kind of high-energy radiation besides cathode rays. On that November day, therefore, he covered a Crookes tube with black cardboard to block out the bright light from the cathode rays. He closed the lab's heavy curtains to make the room as dark as possible. As he expected, he could see no light from the Crookes tube when he turned on its electric current. About a yard (meter) away, however, he detected a faint—and totally unexpected—glow, a tiny cloud of light that seemed to float in the air.

Striking a match, Röntgen found that the glow was coming from a small screen that he had left on a nearby table. The screen was coated with a material that glowed when held next to a source of cathode rays. Röntgen had never heard anyone report, however, that cathode rays could affect such a screen from this great a distance.

When a reporter later asked him what he had thought at this moment, Röntgen replied, "I did not think; I investigated." He turned off the current to the Crookes tube, and the glow disappeared. When he turned the current on again, the faint light came back. He repeated the experiment several more times, each time moving the screen a little farther away from the tube. He continued to see the cloud of light, growing dimmer and dimmer, until the screen was about 2 yards (meters) away from the tube.

Röntgen tried putting a 1,000-page book and then an inch-thick board between the tube and the screen. They had no effect on the glow; the mystery rays seemed to go right through them. These and other tests convinced him that whatever was making the screen glow was not cathode rays. It might be the new kind of radiation he was looking for.

No wonder the professor was not interested in dinner!

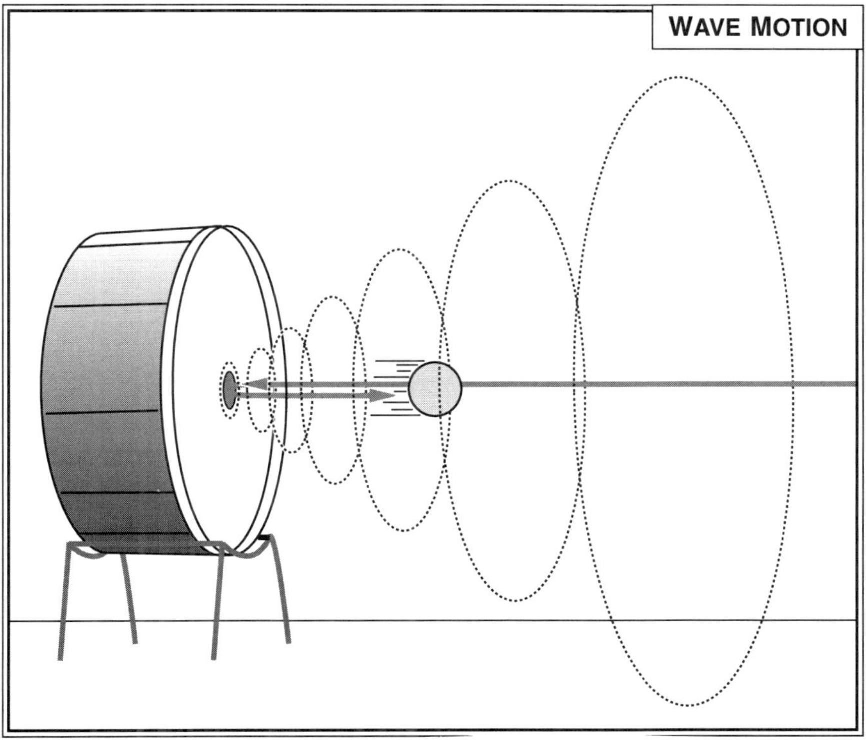

When a ball hits a drum, the ball bounces directly back. It also produces sound waves that spread in all directions. When cathode rays from a Crookes tube (which are actually streams of subatomic particles called electrons) strike a sheet of metal, they bounce back, just as the ball does. They also produce X rays that spread in all directions as the sound waves do.

Thorough Investigation

Wilhelm Conrad Röntgen was born on March 27, 1845, in Lennep, a small town in Prussia, which later became part of Germany. He had taught at two other universities before joining the University of Wurzburg in 1888. He was known as a careful, thorough experimenter.

Röntgen lived up to his reputation in the many tests of his new rays that he made during the next two months. On most days he stayed in his laboratory around the clock. He told no one what he was working on. He said to his coworkers only that he had "discovered something interesting."

Returning to his laboratory after dinner on the night he first found the rays, Röntgen continued testing materials to see which ones stopped the radiation. Among metals, only lead and platinum had much effect.

Far more startling was the discovery Röntgen made when he held a lead weight between the Crookes tube and the glowing screen. To his amazement, he saw on the screen not only a black shadow shaped like the weight but the faint outline of his own thumb and finger. Within the outline the bones of his hand showed clearly, looking almost as dark as the weight. It was as though his flesh had suddenly become transparent, revealing the skeleton beneath.

Röntgen knew that cathode rays could darken photographic film. The new rays, he discovered, had a similar effect. He used the rays in place of light to make several photographs. He found that anything that made a shadow on his screen also showed up in the photos.

On the evening of December 22, Röntgen finally told his puzzled wife what he had been up to. Taking her down to his laboratory, he asked her to put her hand against a sheet of unexposed photographic film. While the patient Bertha waited, Röntgen directed the mysterious rays at her hand for 15 minutes. She then waited again while he developed the film.

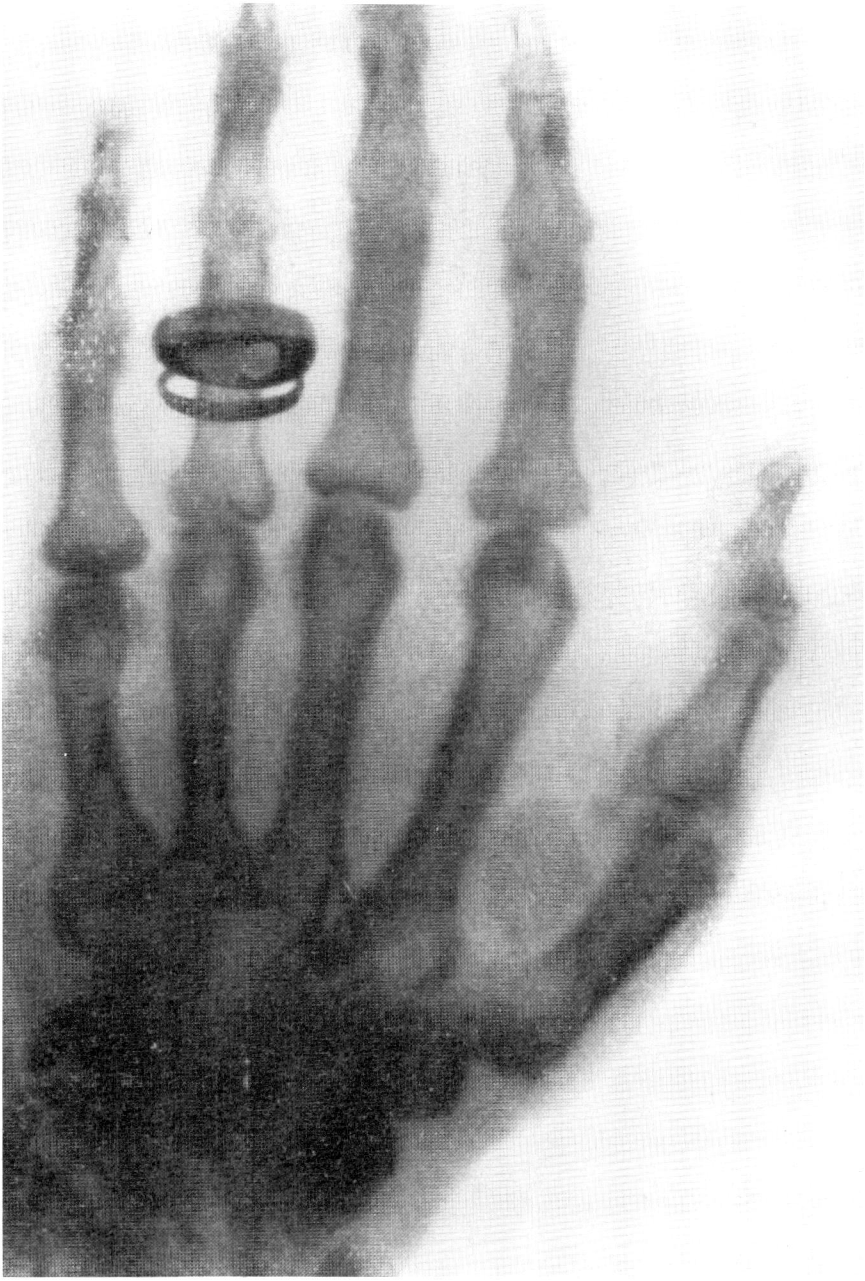

Bertha Röntgen was horrified to see the bones of her hand and her rings revealed by her husband's new rays, as if she had suddenly turned into a skeleton. (Courtesy American Institute of Physics, Emilio Segrè Visual Archives, Lande Collection)

When Bertha saw the finished picture, she gasped in horror. Otto Glasser, a biographer of Röntgen, wrote:

> She could hardly believe that this bony hand was her own and shuddered at the thought that she was seeing her skeleton. To Mrs. Roentgen, as to many others later, this experience gave a vague premonition of death.

"The Devil Will Have to Be Paid"

Röntgen had now completed his first round of tests of the new rays. All his results confirmed that they were not cathode rays or any other familiar kind of radiation. He called them "X" rays, for "unknown." He wrote a paper about them and titled it "A New Kind of Rays."

Röntgen knew that other physicists were also working with cathode rays. They might duplicate his discovery at any time. To show that he had found X rays first, he wanted his paper published as soon as possible. He persuaded Karl Lehmann, president of the

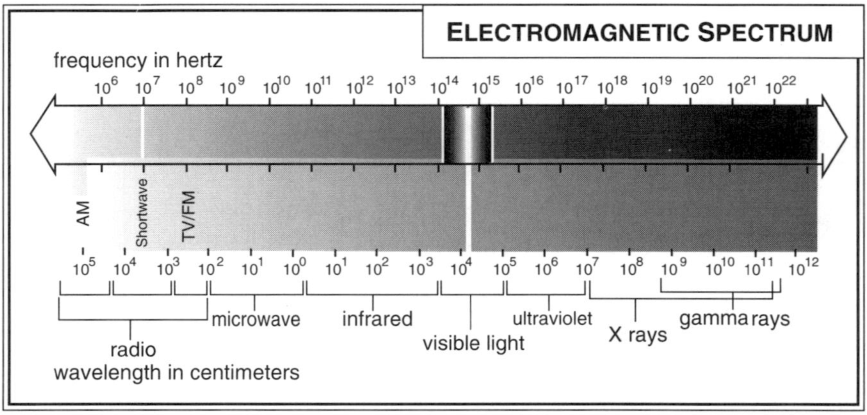

X rays are part of the electromagnetic spectrum of energy. They have higher frequency, shorter wavelength, and greater energy than light waves. This diagram shows where X rays and other familiar forms of radiation fit on the electromagnetic spectrum.

Physical and Medical Society of Wurzburg, to print it in the next issue of the society's journal, which appeared on December 28.

On New Year's Day, Röntgen sent copies of his article and some of his startling photographs, including the one of Bertha's hand, to a number of fellow physicists. As he and Bertha mailed the letters, he made a prophetic remark: "Now the devil will have to be paid."

One physicist Röntgen wrote to lived in Vienna, Austria. He showed the paper to several other Viennese scientists, one of whom was the son of the editor of the newspaper *Vienna Press*. The young man, in turn, showed Röntgen's paper to his father. The editor knew news when he saw it. On January 5, 1896, a Sunday, he put an article about the discovery of X rays on the *Press*'s front page.

The Vienna paper's story was telegraphed around the world. It made headlines in newspapers in London and New York the next day. An American magazine article published a few months later hardly exaggerated when it claimed that "in all the history of scientific discovery there has never been . . . so general, rapid, and dramatic an effect . . . as has followed" the announcement of Röntgen's discovery.

No one felt that impact more than Wilhelm Röntgen himself. Once the Vienna newspaper had blown what Röntgen called "the advertising trumpet," the devil of publicity showed up, just as he had predicted, and demanded a full payment of the professor's time and attention. Röntgen became too busy dealing with reporters and fellow scientists to get back to the work he loved so much. "In a few days I was disgusted with the whole thing," he complained to a friend. "I could not recognize my own work in the reports any more. . . . For exactly four full weeks I have been unable to make a single experiment!"

Röntgen received many honors for his work, including the first Nobel Prize in physics, which he won in 1901. His fame, however, seems to have partly sidetracked his career. He continued to study X rays, but he never made another major discovery. He died in 1923.

I WAS THERE

A VIEW OF THE MYSTERY RAYS

The only reporter who gained a full interview with Röntgen was an American, H. J. W. Dam. In the April 1896 issue of *McClure's Magazine*, Dam described seeing X rays for himself:

> The moment the current passed [through the Crookes tube in Röntgen's laboratory], the [coated] paper began to glow. A yellowish-green light spread all over its surface in clouds, waves, and flashes. The yellow-green luminescence, all the stranger and stronger in the darkness, trembled, wavered, and floated over the paper, in rhythm with the snapping of the [electric] discharge. Through the metal plate, the paper, myself, and the tin box [a portable darkroom, in which Röntgen and Dam stood], the invisible rays were flying, with an effect strange, interesting, and uncanny.

Looking into the Body

Unlike anesthesia and antisepsis, the discovery of X rays caused an immediate stir, not just among scientists, but in the public. Especially in the United States, people were fascinated by science and technology, and they pounced eagerly on popular articles about the new rays. Within a few months, they were flocking to photographic studios to have X-ray photos taken of their hands. Some bought the "Thomas A. Edison X-ray Kit," created by the famed inventor, so they could try X-ray experiments at home.

Doctors, meanwhile, saw at once how X rays could benefit medicine. Many newspaper stories spoke of using the rays to diagnose, or find the cause of, certain kinds of illness by letting doctors see inside the body. For example, a paper in Frankfurt, Germany, pointed out:

> The surgeon . . . could determine the extent of a complicated bone fracture without the manual [by hand] examination which is so painful to the patient. He could find the position of a foreign body, such as a bullet or a piece of shell, much more easily than has been possible heretofore and without any painful examinations with a probe. Such photographs also would be extremely valuable in diagnosing bone diseases . . . and would help to guide the way in therapy [treatment].

McClure's Magazine writer H. J. W. Dam claimed that X rays "appear certainly to be a greater blessing to humanity than even the Listerian antiseptic system of surgery."

ISSUES

THOSE NAUGHTY RAYS

Soon after the discovery of X rays, some people began to be afraid that men would use X-ray glasses to look at women's bodies through their clothes. The New Jersey legislature passed a law forbidding such indecent activity, and a London firm advertised X-ray-proof underclothing. Less seriously, a woman named Wilhelmina wrote in *Photography* magazine:

> I'm full of daze,
> Shock and amaze,
> For now-a-days
> I hear they'll gaze
> Thro' cloak and gown—and even stays
> [undergarments],
> These naughty, naughty Roentgen Rays.

Surgeons put the new rays to work at once. On February 22, 1896, Robert Jones reported in the British medical journal *Lancet* that he had used X rays to locate a bullet in a boy's wrist. This made the task of removing the bullet much easier. Jones's report described the first known use of X rays in diagnosis.

Machines that made X-ray photographs for medical use appeared within a few months of Röntgen's announcement. So did fluoroscopes, devices that projected X rays directly onto a fluorescent screen. If a patient stood or held, say, an arm between the fluoroscope and the screen, an X-ray image of the exposed part of the body appeared on the screen. This let doctors study X-ray images on the spot, without having to wait for photos to be developed.

Some hospitals and doctors bought X-ray machines or built their own. Others sent their patients to studios that specialized in taking X-ray photos. Unfortunately, few of the photographers who set up X-ray studios knew anything about medicine, and few of the doctors who tried to take their own photographs knew

SOCIAL IMPACT

NEW EXPECTATIONS

According to medical historian Donald G. Evens, the success of X rays changed the way Americans and Europeans viewed medicine. People came to expect disease to be cured by new technology rather than by a physician's skill or a change in the patient's habits, such as increased cleanliness. Such expectations increased when antibiotics and other "wonder drugs" were discovered during the 20th century. They continue today, in the era of heart transplants and genetic engineering. Critics feel that this demand for technological miracles has taken attention and funding away from basic research and simpler, less expensive treatments.

CONNECTIONS

X RAYS IN MANY SCIENCES

Medicine was far from the only scientific field that Röntgen's discovery affected. In physics, further study of the rays led within a few years to the discovery of radioactivity, the natural process by which some elements break down into other elements while releasing radiation. Chemists in the 1930s learned how to use X-ray beams to take photographs that showed the structure of complex molecules. This technique, called X-ray crystallography, helped to determine the structure of DNA, the material of which genes are made, in 1953. Today X rays are used as image-making tools in fields ranging from astronomy to electronics.

anything about X rays. No one cared much about that at first. Soon after the turn of the century, however, doctors began to recognize that both kinds of knowledge were needed to take and interpret medical X-ray photographs properly. A new medical specialty, now called radiology, began to be recognized.

Improvements on the first crude machines and techniques came quickly, allowing "X-ray vision" to extend beyond bones. For instance, just after the turn of the century, scientists learned that if they fed people a paste containing the element barium, they could take X-ray pictures of the people's digestive systems. This worked because barium, like lead and bone, stops X rays. Doctors learned how to detect tumors, stones (mineral deposits) in the bladder, and other abnormal conditions with X rays. They also learned much about the normal body by using this new tool.

The idea of using X rays to treat disease appeared almost as quickly as the idea of using them for diagnosis. The first known person to use X rays for this purpose was an Austrian doctor, Leopold Freund. He had read that the hair of an American scientist had fallen out after the scientist had worked with X rays.

In December 1896, therefore, Freund applied the rays to a five-year-old girl who had a large, hairy mole on her back. The treatment worked: the hair sprouting from the mole fell out.

X Rays Show a Dark Side

Over the next several years, doctors tried X-ray treatments on a wide variety of ailments, ranging from acne to cancer. The results were unpredictable and sometimes unpleasant. The back of Freund's little patient, for instance, reddened and blistered as if the child had suffered a bad sunburn. The blisters took months to heal. This setback provided one of the first hints that the new rays could harm as well as heal.

Reddened skin indeed became such a common side effect of X-ray treatment that it was used as one of the first measurements of X-ray dosage. Doctors using the rays for treatment simply exposed people to them until the redness appeared. For many years, there was no more accurate way to measure either the intensity of the rays or their effects on the body. "All one could really do was to place the patient under the [X-ray] machine and hope for the best," James Ewing, one of the first people to use X rays to treat disease, recalled.

X rays could produce much worse effects than a sunburn, though. One of the first people to discover this—the hard way—was Clarence Dally, a glassblower who worked for Thomas Edison. While Dally was trying to invent an X-ray-powered light bulb for the ingenious Edison, his hands were repeatedly exposed to the rays for long periods. Dally's hands turned red, and sores appeared on them. The sores not only refused to heal but spread. Both of his arms finally had to be amputated. Worse still, he developed cancer, almost surely as a result of his X-ray exposure. He died in 1904, the first person known to have been killed by exposure to high-energy radiation. Other people who worked with X rays daily soon began to suffer similar problems, as did sick people treated with high doses of X rays.

In spite of such injuries, the use of medical X rays continued to spread. For instance, both sides in World War I brought portable X-ray machines to battlefields. X rays helped doctors set broken bones and locate bullets or metal fragments in soldiers' bodies.

In the 1920s, a second terrible effect of overexposure to X rays began to appear: Many people who had been exposed to high doses of the rays developed cancer, especially leukemia (cancer of the blood cells). Scientists have since learned that X rays and other forms of high-energy radiation cause cancer by damaging the genetic material in cells. Such cancers often take years to develop.

At the same time, doctors recognized that, oddly, X rays were one of the best treatments for cancer. The rays fight cancer in the same way they cause it—by damaging cells' genetic material. The damage appears only when the cells divide. Cancer cells divide faster than most normal cells, so they are more easily killed by X rays. High-energy radiation is still used to treat cancer.

Advances and Abuses

By mid-century, X-ray machines had improved greatly. The amount of rays that the machines gave out could be controlled more easily. Scientists also could measure X-ray intensity more accurately. Standards for safe levels of exposure to X rays had been set up. Some of the worst abuses of the rays, such as their use in beauty shops to remove unwanted body hair, had been stopped.

Abuses still occurred, however. Safety standards often were not enforced. Machines that were badly made, badly maintained, or improperly used often delivered higher doses of radiation than they were supposed to. X rays were still used for foolish purposes, such as measuring children's feet for new shoes. Diagnostic X rays were overused, too. Many doctors ordered chest X rays as part of their patients' yearly physical examinations whether or not they thought the people might have chest problems, for instance.

Another misuse came from widespread screening programs that were sponsored by well-meaning organizations in the 1950s and early 1960s. In these programs, thousands of people were given chest X rays in the hope of identifying unsuspected cases of tuberculosis, cancer, or other lung diseases. Some groups held contests to see which neighborhood could have the most people screened, and people trying to help their neighborhood went to the screening clinics over and over. These programs found very few new cases of disease, and they greatly increased people's exposure to X rays.

Screening programs like this were discontinued by 1970 because doctors had realized that the risk of damage, and especially of cancer, from X rays was cumulative. Many small doses from things like screening tests and shoe-shop examinations could add up to a lifetime exposure that increased cancer risk significantly.

Doctors also learned that X rays could cause genetic damage when they passed through the reproductive organs. Especially if a woman was pregnant when she was x-rayed, abdominal X rays could increase her children's risk of getting cancer. Because of this

ISSUES

RADIATION FEARS

The public as well as doctors began to question the use of X-ray tests in the 1950s and 1960s. This fear may have arisen partly because of people's concerns about the dangers of radioactive fallout from atomic bomb tests. Hermann Muller, who had won the Nobel Prize in medicine in 1946 for his discovery that X rays caused genetic changes (mutations) in cells, wrote in the mid-1950s, "Our people are annually receiving much more radiation [from medicine] than they do as a result of nuclear test explosions."

TRENDS

INCREASES IN MEDICAL X-RAY EXAMINATIONS

Some critics feel that medical X-ray tests are still overused. Certainly their number has been rising, as these figures for the number of X-ray examinations given in hospitals in the United States show. (Of course, the country's population has been rising, too.)

1964: 58 million
1970: 82 million
1980: 130 million
1995: 300 million (This figure includes both exams given in hospitals and those given outside hospitals.)

danger, X rays are no longer given to pregnant women. People having chest or dental X rays wear lead aprons to shield their reproductive organs.

The greatest improvement in X rays in recent years has been their marriage with computers to create completely new ways of seeing into the body. These new tools will be described in Chapter 7. They have produced a revolution just as exciting as the one that made Bertha Röntgen gasp when, over a hundred years ago, she first saw the bones of her hand lighted by the eerie glow of her husband's "new kind of ray."

Chronology of Röntgen and X Rays

March 27, 1845	Wilhelm Röntgen is born in Lennep, Prussia
1878	William Crookes describes cathode rays
November 8, 1895	Röntgen discovers X rays

December 22	Takes first X-ray photograph of part of human body
December 28	Paper on X rays published
January 5, 1896	First newspaper story about X rays appears
February 22	First use of X rays for medical diagnosis
December	First use of X rays for medical treatment
1901	Röntgen receives first Nobel Prize in physics
1904	First death from effects of X rays
1920s	Large numbers of people exposed to high-dose X rays develop cancer
February 23, 1923	Röntgen dies
early 1970s	Use of X-ray screening programs stopped

Further Reading

Caufield, Catherine. *Multiple Exposures.* London: Secker and Warburg, 1989. Includes interesting material on the popular response to Röntgen's discovery, use of X rays in medicine, realization of dangers presented by X rays, and attempts to control X-ray exposure.

Esterer, Arnulf K. *Discoverer of X-ray: Wilhelm Conrad Roentgen.* New York: Julian Messner, 1968. Biography of Röntgen for young adults.

Evens, Ronald G. "Röntgen Retrospective: One Hundred Years of a Revolutionary Technology." *Journal of the American Medical Association*, September 20, 1995. Review of the use of X rays in medicine and the way X rays changed medicine.

Seliger, Howard H. "Wilhelm Conrad Röntgen and the Glimmer of Light." *Physics Today*, November 1995. Interesting account of Röntgen's discovery of X rays and its background in physics. Difficult reading.

Stroud, Luther, and Neil Sclater. "X-Rays and Their Discoverer." *Electronics Now*, November 1995. Account of Röntgen's discovery, with descriptions of early and modern X-ray equipment and uses.

NOTES

p. 38 "I did not think . . ." Quoted in W. Robert Nitske, *The Life of Wilhelm Conrad Röntgen* (Tucson: University of Arizona Press, 1971), p. 134.

p. 40 "discovered something . . ." Quoted in Nitske, p. 5.

p. 42 "She could hardly . . ." Quoted in Catherine Caufield, *Multiple Exposures* (London: Secker and Warburg, 1989), p. 4.

p. 43 "Now the devil . . ." Quoted in Nitske, p. 98.

p. 43 "in all the . . ." Quoted in Nitske, p. 126.

p. 43 "the advertising trumpet" Quoted in Nitske, p. 100.

p. 43 "In a few days . . ." Quoted in Nitske, p. 100.

p. 44 "The moment . . ." Quoted in Nitske, p. 132.

p. 45 "The surgeon . . ." Quoted in Nitske, p. 114.

p. 45 "appear certainly . . ." Quoted in Nitske, p. 128.

p. 45 "I'm full of daze . . ." Quoted in Nitske, p. 124.

p. 48 "All one could . . ." Quoted in Caufield, p. 10.

p. 50 "Our people are . . ." Quoted in Caufield, p. 146.

4

Blood Is Life

KARL LANDSTEINER AND BLOOD TRANSFUSIONS

Karl Landsteiner, an Austrian scientist, discovered why blood transfusions sometimes succeeded and sometimes failed. (Courtesy Albert and Mary Lasker Foundation and National Library of Medicine)

Blood meant life to Dracula, novelist Bram Stoker's vampire villain. It means life to ordinary people, too. From the beginning of time, anyone could see that a person who lost a lot of blood would die. This vital fluid seemed impossible to replace.

James Blundell, a British doctor, was the first to try to replace one human being's blood with another's. In 1829, he began giving donors' blood to women who had bled heavily during childbirth. The blood was pumped through a tube from an artery in the donor's arm to a vein in the arm of the recipient. Blundell's treatment worked fairly often, but sometimes it failed. No one knew why.

CONNECTIONS

THE FIRST TRANSFUSIONS

Most of the first known blood transfusions, or transfers of blood from one living thing to another, were done not to replace lost blood but to cure insanity. A British doctor named Richard Lower transfused blood from a lamb into Arthur Coga, a young minister, for this purpose in 1667. Coga survived, as did the first two patients of Jean-Baptiste Denis, physician to French king Louis XIV, who performed similar experiments in the same year.

Denis's third patient, a hot-tempered young man whose wife had begged Denis to calm him down, died after three transfusions of blood from a "gentle calf." The ungrateful wife then charged Denis with murder. Denis was acquitted when the court learned that, in fact, the wife had applied her own treatment: a dose of arsenic. Still, this and other deaths threw transfusions into disrepute, and doctors gave up on them for the next 150 years.

KEY CONCEPTS

blood type or group a classification showing which marker chemicals (antigens) are on red blood cells
plasma the liquid part of the blood
shock a condition in which the liquid in the blood is decreased so much that the blood no longer flows
transfusion the transfer of blood from one living thing to another

Different Types of Blood

Just after the start of the 20th century, a scientist named Karl Landsteiner tried to find out why transfusions sometimes failed. Landsteiner had been born in Vienna on June 14, 1868. He was a physician, but he was more interested in doing research than in treating patients.

In 1900, while working at the University of Vienna, Landsteiner found that when blood of different animal species—or animals and humans—was mixed, the mixture separated into what a modern writer has described as "a sea of red dumplings swimming in yellowish liquid." The dumplings are clumps of red cells, the cells that give the blood its color and carry oxygen. If such clumping occurred in the body, severe illness or death would result. This explained why transfusions from animals to humans had seldom worked. The few who had survived such experiments were incredibly lucky.

In a footnote attached to the paper that described this work, Landsteiner mentioned that mixing blood from two human beings sometimes produced the same kind of clumping. He wrote a new paper describing this discovery in more detail in 1901.

Landsteiner tested mixtures of his own blood and that of five other researchers. On the basis of the reactions he saw, he divided human blood into three types, or groups. These were later given the letters A, B, and O (for zero). A fourth type, AB (a combination of A and B), was discovered a year later. Serum (the yellowish fluid) from a person with blood type A would cause clumping in cells of a person with type B blood, and vice versa. Neither would cause clumping in blood of type O. Type O serum would cause clumping in both A and B blood. Type AB serum, on the other hand, did not cause any type of blood to clump.

The serum of the recipient and the blood cells of the donor were what mattered in transfusions, Landsteiner found. Thus type

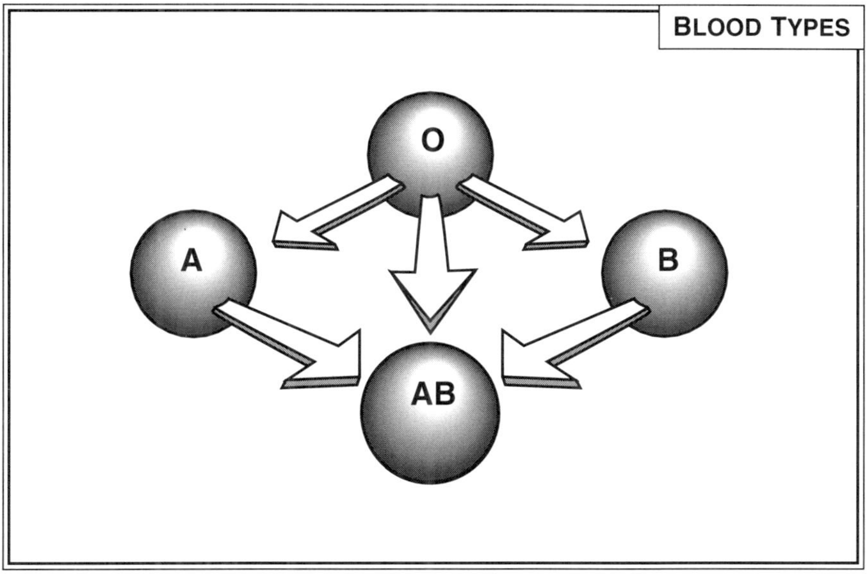

Landsteiner and his coworkers discovered that each person belongs to one of four main blood types or groups: A, B, AB, or O. Anyone can safely give blood to or receive blood from anyone else belonging to the same group. People belonging to group O can give blood to anyone but can receive blood only from other type O people. People of types A and B can give blood to people of type AB but not to each other. People of type AB can receive blood from anyone but can give it only to other type AB people.

O blood, which did not clump when mixed with any kind of serum, could be given to people of any blood type. People with blood types A and B, however, could give blood only to people of their own blood type or to those with type AB. Type AB people could receive blood from anyone but could give blood only to other AB people.

These reactions are caused by the immune system, the combination of cells (mostly in the blood) and chemicals that protect the body against invasion by disease-causing microorganisms and other foreign substances. Certain immune cells in the blood identify other cells as foreign by detecting chemical markers on the cell surfaces. The immune cells then make other chemicals called antibodies, whose structure is tailored to match the foreign markers. These antibodies float in the blood serum. If they encounter cells with the markers they match, they attach to the markers as a key fits into a lock. Their presence on a cell sends other immune system cells a message that says in effect, "Attack this cell—it doesn't belong here." Those cells then destroy the invading cell. The clumping that Landsteiner saw was a sign of this destruction.

The markers on the cell surface to which antibodies react are called antigens. Landsteiner showed that people with type A blood have one kind of antigen on their red blood cells. The red cells of people with type B blood carry a different antigen. People with type O blood have neither antigen. Those with type AB blood have some cells with A antigen and some with B. People have antibodies only to the antigens that their own blood cells do not carry. Thus, antibodies from a person with type A blood will mark cells with the B antigen for destruction and vice versa.

Landsteiner's work explained why transfusions between human beings sometimes succeeded and sometimes failed. If donor and recipient of a transfusion had incompatible blood types, antibodies in the recipient's serum would make the red cells of the donor's blood clump in the recipient's body. The clumps would block blood vessels, cutting off blood flow.

Using Landsteiner's discovery, technicians could easily find out ahead of time whether a transfusion would save a life or take one.

CONNECTIONS

BLOOD DETECTIVES

Blood typing, or identifying the blood group to which a sample of blood belongs, can reveal much more than whether a transfusion will work. As the O. J. Simpson "trial of the century" showed, scientists in police labs use typing to find out whether blood found at the scene of a crime could have come from a particular victim or suspect. Typing can also determine whether a man could be the father of a certain child. Because typing is much more complex than it was in Landsteiner's day—14 different sets of blood groups are now known—a complete description of a person's blood type is almost as unique as a fingerprint.

They just mixed a few drops of blood from the potential donor and recipient and watched to see whether clumping occurred. If it didn't, the transfusion probably would be successful.

Neglected Discoveries

Landsteiner's discovery of blood types should have caused as much excitement as the discovery of X rays did. After all, people died from loss of blood every day. Others died from a condition called shock. Transfusions could have saved the lives of both groups.

Shock often follows major loss of blood or severe injury, such as widespread burns. Victims of shock turn pale and cold, break out in a clammy sweat, and often lose consciousness. Their pulse can hardly be felt. Doctors learned that in shock, the amount of liquid in the blood vessels falls so low that the blood can no longer flow. Adding fresh blood to the body thus ought to help people in shock as well as those who had lost blood.

In 1905, a Cleveland surgeon named George Crile did try using blood transfusions to treat shock. He performed Landsteiner's typing test to make sure that the donor's and recipient's blood would be compatible. He then transfused the blood directly from donor to recipient, just as Blundell had done. Crile's treatment usually succeeded, but few other doctors followed his lead.

Probably the main reason why most doctors did not start using blood transfusions right away was that no one knew how to preserve blood. Once removed from the body, it quickly formed a gooey mass called a clot, just as it does when it flows out of a wound. (Clotting is different from the clumping of cells that occurs when incompatible blood types are mixed.) Transfusions thus had to be made on the spot, with donor and recipient next to each other. Finding a compatible donor quickly enough to save a life in an emergency was almost impossible.

In 1914, just about the time World War I began, several scientists in different parts of the world independently made what should have been another breakthrough discovery. They found that if a chemical called sodium citrate was added to fresh blood, clotting was prevented for about 10 days. A few surgeons gave transfusions of citrated blood to soldiers suffering from blood loss or shock during the war. Most battlefield doctors, however, preferred to use salt solution rather than blood to treat shock. By adding liquid to the body's remaining blood, the solution kept the blood flowing, and, unlike blood, it did not have to be typed.

Monkeys and Babies

Karl Landsteiner moved to the United States in 1922 and went to work for the Rockefeller Institute for Medical Research in New York. In 1930, he finally received the recognition he deserved: He was awarded the Nobel Prize in medicine for his discovery of the A, B, and O blood types.

Landsteiner made a second major discovery in 1940. Working with Brooklyn scientist Alexander Wiener, he found that rabbits could make antibodies to the blood of rhesus monkeys, a type of Asian monkey commonly used in laboratory experiments. To their surprise, the scientists discovered that these rabbit antibodies also clumped the cells of about 85 percent of the New Yorkers they tested. Monkeys and most humans, it seemed, had a red cell

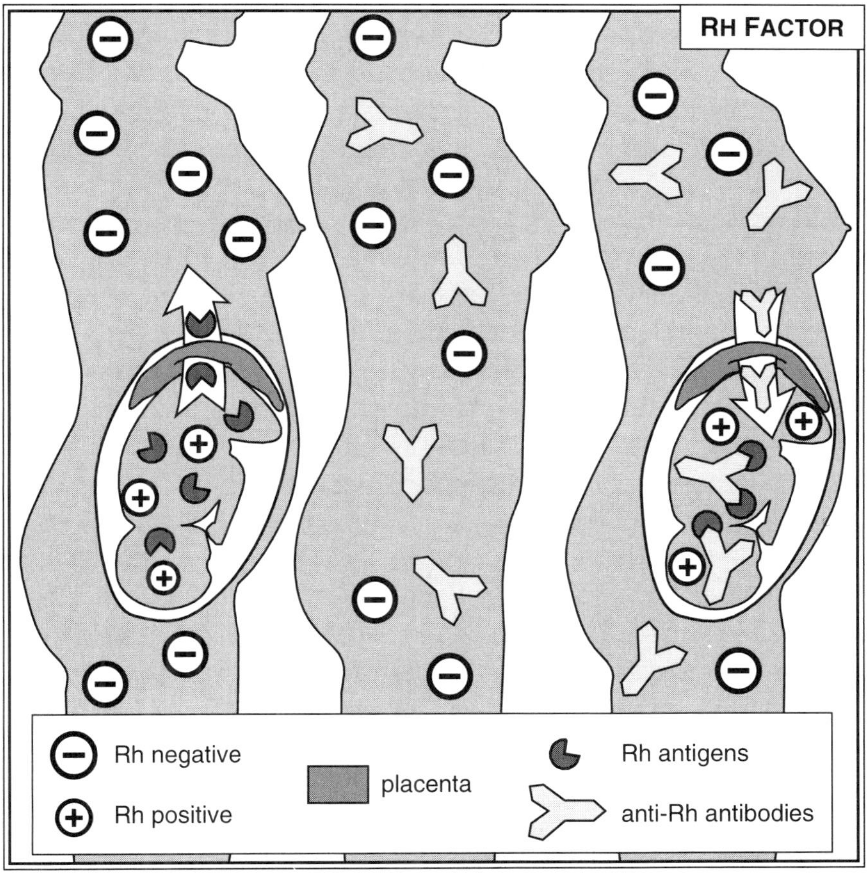

If a woman whose blood lacks the Rh factor (is Rh Negative) carries a child whose blood contains this factor (is Rh Positive), some of the unborn baby's Rh antigens may seep into her bloodstream before or during birth. Her immune system makes antibodies to attack these "foreign" chemicals. If she later becomes pregnant with another Rh-positive child, her antibodies may enter its circulation and destroy its blood cells.

antigen in common. Landsteiner called this antigen the rhesus, or Rh, factor. The finding of the Rh factor was Landsteiner's last important discovery; three years later, he died of a heart attack.

The discovery of the Rh factor explained why transfusions sometimes made people sick even when donor and recipient had compatible ABO blood types. When the blood of a person whose cells had the Rh factor—in Landsteiner's terms, someone who was Rh positive—was given to someone who lacked the factor, or was Rh negative, the Rh-negative person formed antibodies to the Rh antigen. If that person ever received a second transfusion of Rh-positive blood, the antibodies attacked and destroyed the cells of the donated blood.

The Rh factor also explained a medical mystery that had nothing to do with transfusions. Sometimes women gave birth to dead or very sickly babies for no obvious reason. Researchers now noticed that these women were often Rh negative and were married to men who were Rh positive. Small amounts of a baby's blood can enter the mother's circulation before or during birth. If an Rh-negative woman conceived a child who was Rh positive, like its father, she would form antibodies to the Rh antigen in its blood. These antibodies caused no harm to that child, but if the woman later had another Rh-positive child, her antibodies would attack its red blood cells.

Today, if an Rh-negative woman gives birth to an Rh-positive child, she is given anti-Rh antibodies right afterward. These short-lived antibodies destroy any of the baby's cells that have gotten into the mother's blood before her immune system has time to form its own longer-lasting antibodies.

Banking Blood and Plasma

By the time Landsteiner discovered the Rh factor, blood transfusions had finally begun to be used on a large scale. In 1933, a Russian scientist named Sergei Yudin found that blood treated with sodium citrate could be kept for a month or two if it was refrigerated. Bernard Fantus, a Chicago doctor, used this technique in 1937 to create what he called a blood bank. Blood from donors was typed ahead of time and stored in the bank's refrigerators. It was then ready to use when needed.

Whole blood was not the only thing stored in blood banks. Beginning soon after World War I, doctors had realized that for many purposes, including treatment for shock, blood plasma was as useful as whole blood. Plasma is the liquid part left after cells are removed from blood. Like salt solution, plasma restores liquid to blood, but it does so more effectively because of chemicals it contains.

Using plasma rather than whole blood solved several problems. Plasma contained no cells, so plasma from any donor could be given to any recipient. Plasma could be kept in refrigerated bottles for up to two years. Better still, researchers discovered in 1935 that plasma could be frozen, dried in a vacuum, and stored as a powder. A small bottle of plasma flakes, a pint of sterile water, tubing, and a hypodermic needle made up a "plasma kit" that was easy to ship. When the plasma was needed, the flakes were mixed with the water and injected.

These advances came just in time. With Adolf Hitler and his warlike National Socialist party (the Nazis) in power in Germany, many people in Europe and America suspected that Europe, and perhaps the whole world, would soon be at war. Having blood in a form that could be preserved, stored, and shipped easily was likely to become vital.

Blood Goes to War

In 1939, the Germans began invading nearby countries, and the predicted war in Europe started. German planes bombed British cities, producing many casualties, and Britain needed plasma urgently. The United States, not yet at war itself but wanting to help Britain, began a program called "Blood for Britain" in June 1940. During the five months of its operation, the program shipped more than 17,000 pints (7,990 l) of plasma to the

African-American surgeon Charles Richard Drew, shown at left, headed national blood and plasma collection programs in the United States around the start of World War II. Among other things, he created the Bloodmobile, or mobile blood collection van. (Courtesy Moorland-Spingarn Research Center, Howard University)

beleaguered country. American plasma saved British lives until Britain set up its own blood banks.

The head of the Blood for Britain program was a young African-American surgeon named Charles Richard Drew. Drew had done research on transfusions and methods of preserving blood, and he was also skilled at organizing people. He set up a system for collecting blood from volunteers, extracting plasma from it, and storing and shipping the plasma. He made sure that all the blood banks contributing to the program followed the procedures and strict standards he set.

Drew took on a new responsibility when the Blood for Britain program ended. Suspecting that the United States would soon be at war, the armed forces asked the American Red Cross to set up a national blood bank system. Drew was made medical director of this program early in 1941. By the time the United States entered the war on December 7, 1941, the program had prepared a million pints (470,000 l) of plasma.

During World War II, blood collection drives in the United States brought in over 13,000,000 pints (6,110,000 l) of blood. Blood of type O, the "universal donor" type, was kept for use in whole blood transfusions. The rest was turned into plasma.

Plasma traveled to battlefields everywhere. It was dropped from planes and carried through jungles on the backs of burros. Soldiers so badly hurt that they seemed almost dead received plasma transfusions and a few minutes later, according to one wartime report, "would be sitting up and talking, with all the life and color back in their faces." With their blood volume restored, they could survive long enough to reach more extensive medical help. Partly because of the new availability of blood and plasma, the death rate

SOCIAL IMPACT

THE COLOR OF BLOOD

When the United States first began stockpiling blood and plasma for possible war use, it refused to accept blood from African Americans. This understandably made Charles Drew very angry. At a press conference he pointed out, "The blood of individual human beings may differ by groupings [types], but there is absolutely no scientific basis to indicate any difference according to race." There was thus no reason why safe transfusions between races could not occur. Drew's logic, unfortunately, was no match for prejudice. Once the war began, the army started accepting African-American blood, but it gave the blood only to black soldiers.

OTHER INVENTORS

EDWIN COHN AND ALBUMIN

One problem with dried plasma was that it had to be mixed with sterile water before it was transfused. The mixing could take 10 minutes, time that was not always available on a battlefield. Fortunately, Edwin Cohn, a professor of biochemistry at Harvard Medical School, discovered in 1940 that one chemical in plasma, albumin, could do most of the work of the plasma. When injected into the bloodstream, albumin absorbs liquid from the surrounding tissues and thus restores blood volume. A bottle of albumin the size of a fist could do the work of a plasma kit 10 times that size. Albumin, like plasma, became a battlefield lifesaver.

among wounded soldiers in World War II was less than half that in World War I.

A Gift of Life

Blood banks are still an essential part of most countries' health care systems. People in the United States donate about 12 million pints (5,640,000 l) of blood each year to some 2,400 blood banks across the nation.

Opposite: *A pint of blood goes on a long journey from donor to recipient. (1) Blood drives bring in people who are willing to help others by giving their blood. (2) Donors are asked questions to see if they are likely to be carrying a blood-borne disease. (3) If they give the right answers, they are allowed to donate blood, a safe and almost painless process that takes about 10 minutes. (4) The blood is tested to see whether it contains any signs of blood-borne diseases. (5) The blood is separated into as many as 20 parts or fractions, each with a different medical use. (6) Blood banks send most of their blood to local hospitals, but if they have extra supplies, the blood may go farther away. (7) Transfusions of blood or blood products are given to those who need them, usually in hospitals.*

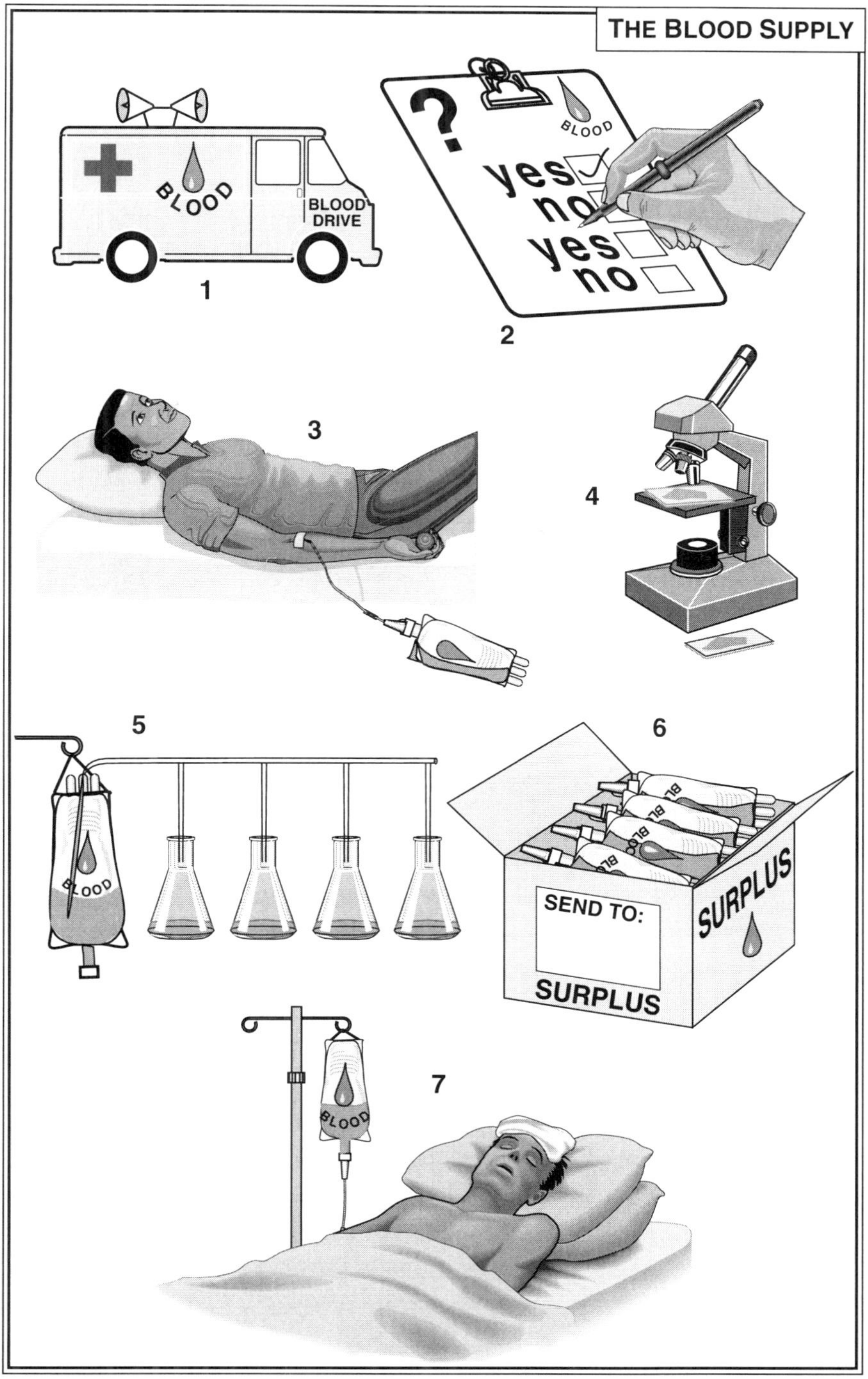
THE BLOOD SUPPLY
BLOOD
BLOOD
DRIVE
1
BLOOD
yes
no
yes
no
2
3
4
5
BLOOD
6
SEND TO:
SURPLUS
SURPLUS
7
BLOOD

Donating blood is easy to do, causes no pain except for the prick of a needle, and presents no health risk. It takes only 10 to 20 minutes. Typically, a donor first receives a brief physical examination and answers some questions about his or her health and lifestyle. These questions help to determine whether the person is likely to carry any disease, such as AIDS, that is spread through blood. If the answers are satisfactory, a nurse pricks the person's finger and withdraws a drop of blood. This blood is tested quickly to see if it contains enough iron and certain other chemicals.

If the blood passes the test, the donor lies back in a comfortable chair while a nurse puts a hypodermic needle into an arm vein. Blood flows out through a tube and is collected in a sterile plastic bag. Two test tubes full of blood are also withdrawn for laboratory tests. The bag and tubes are labeled with a number so they can be easily identified.

Some blood in blood banks is grouped by blood type and stored in large refrigerators. It can be kept from 21 to 42 days. Less than 2 percent of blood collected by blood banks is stored or used as

ISSUES

HOW SAFE IS OUR BLOOD SUPPLY?

The largest blood banks, which take in about 90 percent of the blood donated in the United States, are licensed by the federal Food and Drug Administration (FDA) and must follow its strict standards. However, critics such as journalist Andrea Rock, writing in *Money* magazine in 1994, maintain that "the ways in which crucial decisions [about blood banking] are made [by the government] remain biased toward the blood banking industry's business interests." Similarly, a 1994 article in *U.S. News and World Report* claims, "Transfusions are far riskier than patients are led to believe." That article says that about 1,000 people with HIV and 46,000 with hepatitis manage to give blood each year.

TRENDS

BLOOD UNIT RECALLS

Units of blood are recalled, or withdrawn from the banking system, if a donor's health problems are uncovered after donation. *U.S. News and World Report* gives the following numbers of units of blood recalled in recent years:

1990: 26,000
1991: 309,000
1992: 21,000
1993: 11,000

whole blood, however. The rest is divided into as many as 20 parts or fractions.

Each blood fraction has a different use. Red cells are given to people who have lost blood and must have the oxygen-carrying power of their blood system restored. Plasma and albumin restore the blood's liquid volume, just as they did during World War II. Platelets, small bodies that help the blood clot, are often given to cancer patients whose own platelets have been damaged by anti-cancer treatments.

While the main part of the donated blood is being separated into fractions, the blood in the test tubes is tested for signs of eight blood-borne diseases. These include AIDS and hepatitis, a serious liver disease caused by a virus. Because of these tests, chances of catching diseases from blood transfusions in the United States are small.

Because receiving a transfusion presents some health risk, people who know they are going to have surgery often donate blood ahead of time for their own use. People can also request that blood shed during their surgery be saved and returned to their bodies.

In spite of the small risks involved, some 3.6 million Americans receive transfusions of blood or blood products every year. Those transfusions have often saved their lives.

Chronology of Blood Transfusion Discoveries

1667	First transfusions from animals to humans
1829	James Blundell makes first human-to-human transfusion
June 14, 1868	Karl Landsteiner is born in Vienna
1900–1901	Landsteiner publishes papers that describe human blood types (A, B, O, and AB)
1905	George Crile treats shock with blood transfusions
1914	First use of sodium citrate to preserve blood
1930	Landsteiner receives Nobel Prize in medicine
1933	Sergei Yudin uses refrigeration to preserve blood
1937	Bernard Fantus sets up first blood bank
1940	Landsteiner and Wiener discover Rh (rhesus) factor "Blood for Britain" program provides mass shipments of plasma to Britain Edwin Cohn extracts albumin from blood
1940–1941	Charles Drew directs U.S. blood bank programs
1943	Landsteiner dies
1985	Donated blood first tested for AIDS

Further Reading

Blood: The River of Life. New York: Torstar Books, 1985. Well illustrated; contains much interesting material on blood, including the development of blood transfusions.

Dixon, Bernard. "Of Different Bloods." *Science 84*, November 1984. Describes Karl Landsteiner's discovery of blood types and several modern uses of blood typing.

Lee, Sally. *Donor Banks*. New York: Franklin Watts, 1988. For young adults. Has chapter on blood banks.

Riedman, Sarah R. *Masters of the Scalpel*. Chicago: Rand McNally, 1964. Has chapter on development of blood transfusions and prevention of blood loss in surgery.

Starr, Douglas. "Again and Again in World War II, Blood Made the Difference." *Smithsonian*, March 1995. Describes use of blood, plasma, and albumin to save lives during World War II, focusing on contributions of Edwin Cohn.

Yount, Lisa. *American Profiles: Black Scientists*. New York: Facts On File, 1991. For young adults. Contains chapter on Charles Drew and his work in blood bank programs.

NOTES

p. 55 "gentle calf" Quoted in *Blood: The River of Life* (New York: Torstar Books, 1985), p. 44.

p. 56 "A sea of . . ." Jack Mingo, "The Business of Blood," *East Bay Express*, January 24, 1997, p. 9.

p. 65 "would be sitting up . . ." Quoted in Douglas Starr, "Again and Again in World War II, Blood Made the Difference." *Smithsonian*, March 1995, p. 135.

p. 65 "The blood . . ." Quoted in Lisa Yount, *Black Scientists* (New York: Facts On File, 1991), pp. 61–62.

p. 68 “the ways in which . . .” Andrea Rock, “America’s Dangerous Blood Supply,” *Money*, May 1994, p. 136.

p. 68 “Transfusions are far . . .” Richard J. Newman and Doug Podolsky with Penny Loeb, “Bad Blood,” *U.S. News and World Report*, June 27, 1994, p. 69.

5

From Sausage Casings to Six-Million-Dollar People

WILLEM KOLFF, ROBERT JARVIK, AND ARTIFICIAL ORGANS

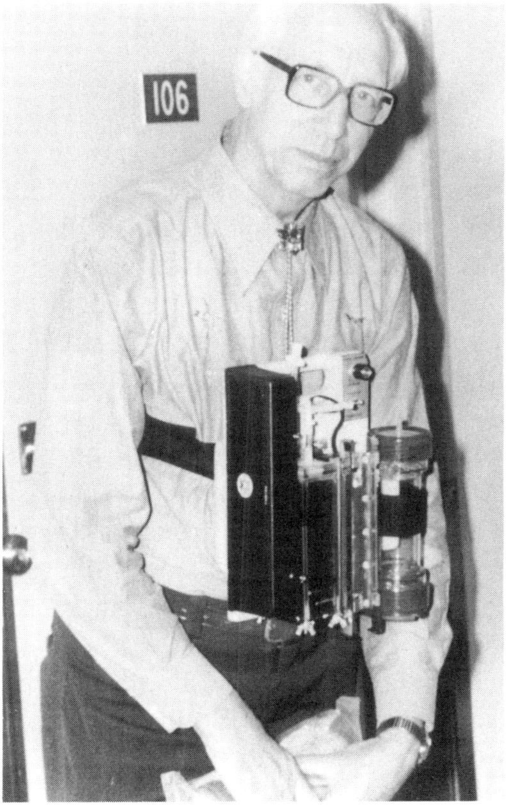

Dutch physician and inventor Willem Kolff has been called "the father of artificial organs." He designed the artificial kidney (dialysis machine), the first artificial organ, and also worked on artificial hearts and other organs. Here he is wearing a portable form of artificial kidney. (Courtesy Manuscripts Division, University of Utah Libraries)

Willem Johan Kolff almost didn't become a doctor because he didn't think he could stand to see people die. And indeed, when the young Dutch physician (born on February 14, 1911, in Leyden) watched a man die of kidney failure in 1938, he hated it so much that he made a very important decision. The decision was not to quit medicine, however. Instead, Kolff decided to try to invent a machine that could save the lives of people like this man by doing the kidneys' work.

The kidneys are two bean-shaped organs in the lower back. Their chief job is removing wastes from the blood. The wastes are sent out of the body in the form of urine. If both kidneys are damaged by injury, poison, or disease, toxic waste products build up. A person can die of kidney failure within a few days.

Building a Filter

Kolff knew that mixed substances could be separated by the process of filtering. A filter is a thin sheet of material, or membrane, with microscopic holes in it. Particles smaller than the holes pass through the filter, while larger particles remain behind. When a filter separates substances in a liquid, allowing the smaller particles to flow into a second liquid on the other side of the filter, the process is called dialysis.

The kidneys are living filters. Kolff thought he might be able to design a device that would perform the same function, at least in a crude way. He knew that as early as 1913, researchers at Johns Hopkins University in the United States had made a dialysis machine that worked on dogs. No one had ever tried an artificial kidney on a human, though.

Kolff knew of one cheap, easy-to-find material that he might use as a filter: a stiff, transparent plastic called cellophane. At that time it was used to make casings for sausages. In 1939, he poured a mixture of blood and urea (one of the chief compounds the

KEY CONCEPTS

bionics use of engineering techniques to make artificial devices that substitute for parts of the body; also called biomedical engineering or bioengineering

dialysis the process of separating substances in a solution by filtering them through a membrane

stroke sudden damage to a blood vessel in the brain

ventricle one of the lower two of the four chambers of the heart; the ventricles pump blood into the body

kidneys must filter out of the blood) into a cellophane sausage casing, sealed the casing, and rocked it back and forth in a container of salt water. After half an hour, the urea had moved out into the water, and the blood in the casing was clean.

Using only materials that were inexpensive and easy to get was important because soon after Kolff began his research, Holland was forced into war. Its neighbor, Germany, invaded and took control of it in 1940. German Nazis and their Dutch supporters controlled supplies of all valuable materials.

Like most of his countrymen and -women, Kolff resisted the Nazis in any way he could. Rather than work under a Nazi doctor, he abandoned his job in the city of Groningen and moved to a smaller town called Kampen. There he sometimes risked his life by hiding anti-Nazi underground members in the hospital where he worked.

In Kampen, Kolff spent every spare moment trying to turn his sausage-casing filter into something that could be used on a human being. He scrounged materials anywhere he could find them. For instance, he got packing to stop leaks around the tubes in his device from a nearby Ford dealer. The material had been used to seal the water pumps in the dealer's car engines.

The First Artificial Kidneys

Kolff first used an artificial kidney on a human patient on March 17, 1943. The machine, about the size of a baby carriage, was a wooden drum wrapped with many feet of cellophane tubing. It sat in a tub partly filled with a liquid called dialysis fluid, which contained certain chemicals that the body needs.

One end of the tubing was connected to an artery in the patient's arm, the other end to a vein. The blood flowed out of the artery and into the cellophane tube. As an electric motor slowly turned the drum and the blood traveled through the tubing, urea and other wastes passed through the cellophane into the dialysis fluid. Useful chemicals passed from the fluid into the blood at the same time. The blood finally went back into the patient's body through the vein.

Kolff's first patient, a 29-year-old woman, died after 12 dialysis treatments. All but 1 of the next 14 people Kolff tried his invention on died quickly, too—and the lone survivor, he knew, might well have lived without any treatment. Kolff did not feel that he had failed, however. He could see that his machine had made a difference in what remained of these very sick people's lives. He said later:

> I saw them regain their consciousness. I saw them talk to their families. I saw them read the newspapers, write their wills. Even when I lost them, . . . I knew that I had seen a temporary improvement. I was sure that in time I would get one who would be saved.

Kolff continued to work throughout the war, slowly improving his machine. Then finally, in September 1945, a few months after the Germans had been driven out of Holland, Kolff was able to claim a victory. His artificial kidney kept a 67-year-old woman alive long enough for her own kidneys to recover from their sudden failure, and she was restored to health. Ironically, this woman, Sofia Schafstadt, had been a Dutch Nazi. She was in prison, awaiting trial for war crimes, when her kidneys failed.

Willem Kolff's artificial kidney, or dialysis machine, was the first device that successfully replaced a major human organ. With its invention, Kolff founded the science of bionics, or biomedical engineering. He has been called "the father of artificial organs."

From Kidneys to Hearts

Kolff and his family moved to the United States in 1950. He began working at the Cleveland Clinic in Ohio, where he set up the world's first artificial organ research center. At first he continued to design better dialysis machines. His original machines had been intended only for short-term use, keeping people alive until their own kidneys could function again. Around 1960, however, researchers figured out ways to use dialysis machines on people with permanent kidney failure. Such people had to go to a hospital several times a week and spend up to 16 hours hooked up to the machines.

At first few hospitals had dialysis machines, and dialysis was very expensive. Kolff wanted to make an artificial kidney that anyone could afford. He also hoped to create a machine that people could use at home. In 1966, he and a coworker, Yukihiko Nose, began making one model of a cheaper kidney from, of all things, washing machines. After a while, though, the washing machine manufacturer, fearing a lawsuit if one of Kolff's devices failed, refused to sell him any more machines. "It just shows you how incredibly narrow-minded people can be," Kolff said later.

Greatly improved models of dialysis machines are still saving lives. Some dialysis patients must go to a hospital for treatment, but others have machines that they can use at home. Some forms of dialysis machine are wearable and allow their users to lead more

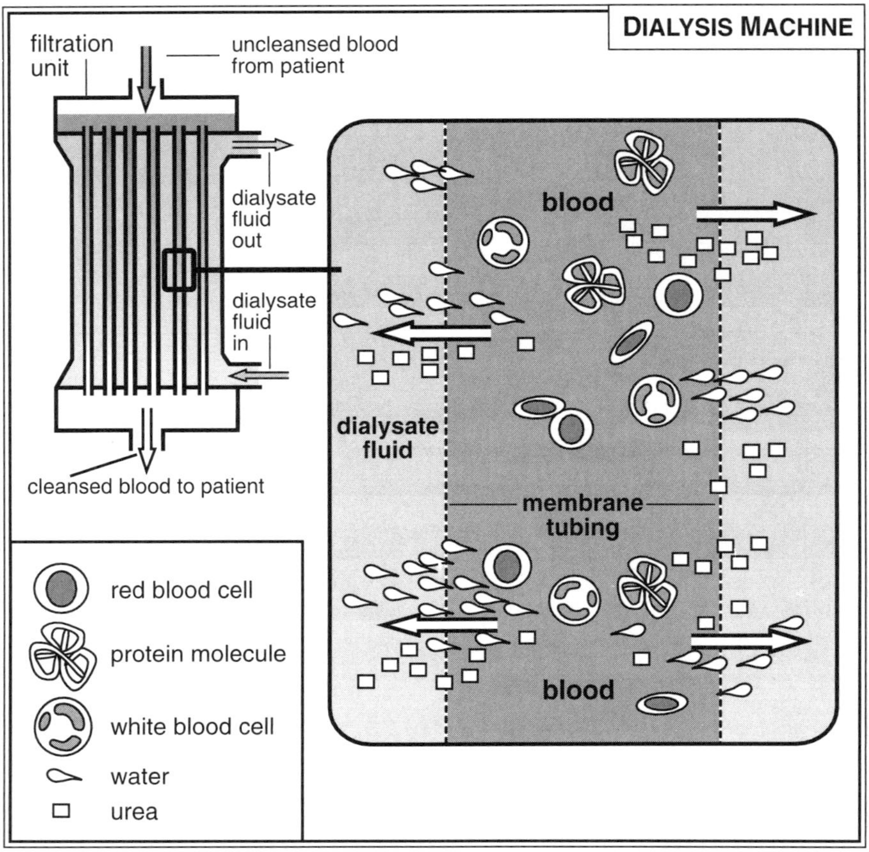

The artificial kidney, or dialysis machine, performs the kidneys' main job of filtering wastes from the blood. The blood is taken from an artery in a person's arm, passed through the machine, then returned to an arm vein. Inside the machine, the blood flows through thin tubing made of a material that contains microscopic holes (dotted lines). Blood cells and complex molecules remain inside the membrane and eventually go back into the body. Molecules of urea and other wastes, along with some water, pass through the membrane into a fluid called dialysis fluid, which is later discarded.

or less normal lives. Dialysis devices now are used mainly to keep people alive until they can get a kidney transplant (see Chapter 6).

As well as improving dialysis machines, Kolff started work on artificial hearts. Machines that could take over the work of the heart and lungs during surgery had existed since 1953, and at first Kolff made improvements in this type of machine. His long-term goal, though, was a heart that could remain permanently in the body. He implanted his first artificial heart into a dog in 1957. That pioneering dog lived for 90 minutes.

Kolff continued his work on both kidneys and hearts when he moved to the University of Utah, in Salt Lake City, in 1967. The university's medical center made him the director of its Division of Artificial Organs and its new Institute of Biomedical Engineering.

CONNECTIONS

JOHN H. GIBBON AND THE HEART-LUNG MACHINE

The heart-lung machine, like the artificial kidney, was invented because a doctor could not stand to see a patient die. In 1930, a young surgeon named John H. Gibbon, Jr., assisted in an operation to remove a clot that was blocking a major blood vessel near a woman's heart. The brain can survive without oxygen from the blood for just six minutes. This operation took seven minutes, and the woman died.

Spurred by this tragedy, Gibbon and his wife, Mary, began work on a machine that could take over the function of the heart and lungs during surgery. A pump substituted for the heart, sending blood from the body into the machine. Inside the machine, the blood was spread in a thin film on a wire screen in a turning metal cylinder. Oxygen was blown onto this film. Then the blood, now carrying a load of precious oxygen, was pumped back into the body. With its work being done by the machine, the heart could be stopped for hours. Heart transplants, implantation of artificial hearts, and all types of open heart surgery depend on the descendants of Gibbon's heart-lung machine.

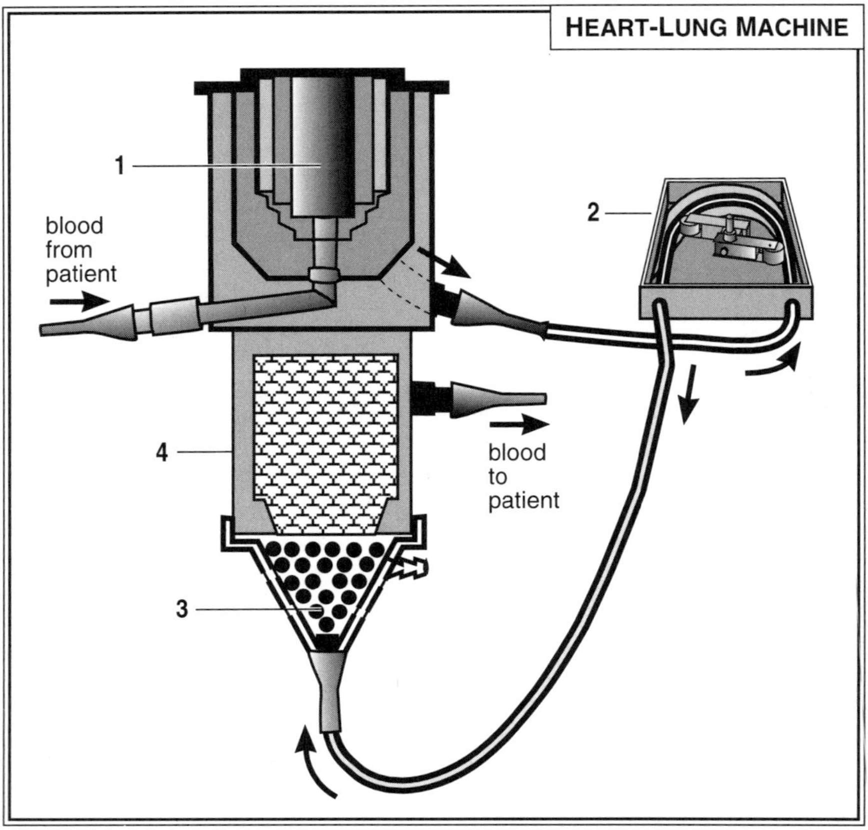

The heart-lung machine does the heart's job of pumping blood and the lungs' job of adding oxygen to the blood, allowing the heart to be stopped for long periods during surgery. Blood from the patient's body flows into a reservoir (1), where it is cleaned. It then passes through a pump (2), which sends it to a heat exchanger (3) for cooling. Cooling the blood slows the patient's body processes and reduces the need for oxygen. Finally, the blood goes into an oxygenator (4), where oxygen is added to it. It then is pumped back into the patient's body.

The Jarvik Heart

Almost as important to bionics as Willem Kolff's inventions is his skill at attracting and training young people as talented as he is. "I take enthusiastic young people from different fields and bring them together to work on imaginative projects," he says. "That's probably what I'm best at." Many of the best inventors in bionics have worked under him. Two brilliant young men who went to work for Kolff soon after he moved to the University of Utah were William C. de Vries, a surgeon, and Robert Koffler Jarvik. Both, along with Kolff himself, would play key roles in the development of the first permanent artificial heart.

When Robert Jarvik joined Kolff in 1971, he had not yet gotten his medical degree. Born in Midland, Michigan, on May 11, 1946, Jarvik had changed his career plans several times. He had always been interested in medicine—he even invented a device to improve

Working in Willem Kolff's laboratory at the University of Utah, Robert Koffler Jarvik designed the first permanent artificial heart implanted in a human being. (Courtesy Manuscripts Division, University of Utah Libraries)

suturing, or stitching, in surgery while still in high school— but determined to become a doctor only after his father developed heart disease. ("I decided I better do something about this," he said later.) Jarvik earned a master's degree in bioengineering from New York University and finally finished his medical training at the University of Utah in 1976 while working with Kolff.

In Kolff's laboratory, Jarvik began working on the artificial heart. Like Kolff's artificial kidney, the heart was a greatly simplified

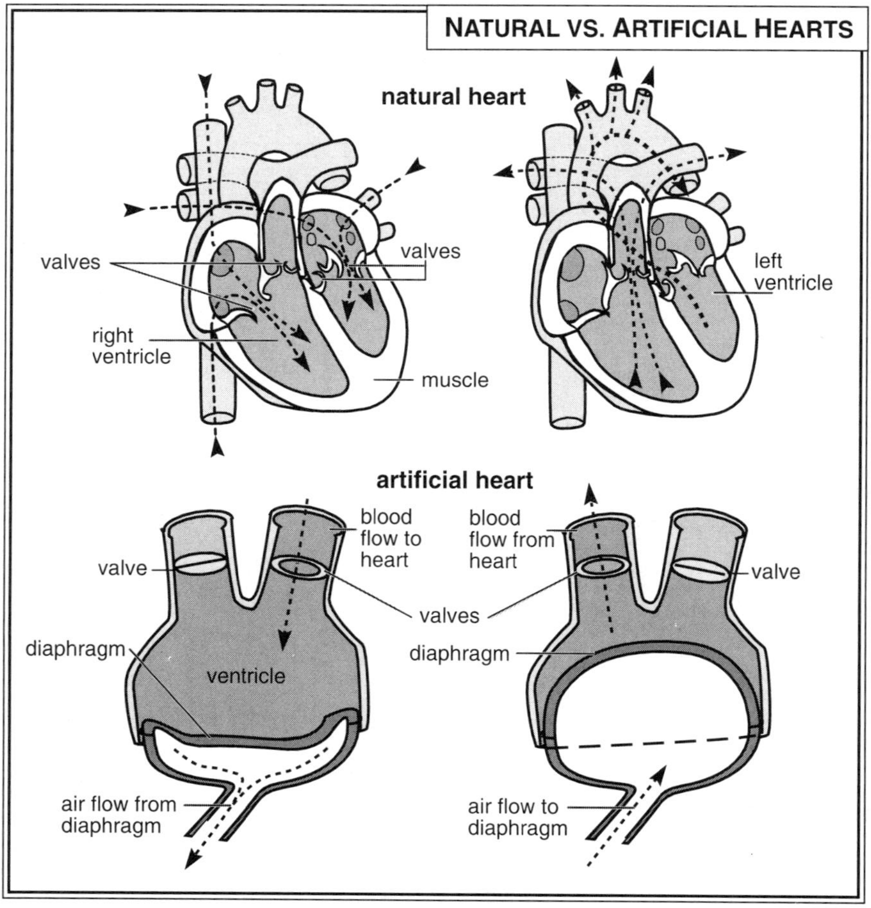

These pictures compare the natural heart (top) and the Jarvik-7 artificial heart (bottom), showing similarities and differences in their structure and in the way blood flows into and out of them.

SOLVING PROBLEMS

HOW TO MAKE A HEART

Kolff's group made the first diaphragms for their artificial hearts from silicone rubber. Unfortunately, blood clots tended to form on the diaphragms' surfaces. Robert Jarvik's first contribution to the artificial heart was his suggestion to make the diaphragms of smooth, elastic polyurethane instead. This form of the plastic is similar to Lycra, an elastic fabric used in women's swimsuits and underclothing.

version of the organ it replaced. Just as a kidney is basically a filter, a heart is basically a pump. Its job is to move blood through the body. In a human being, most of the pumping work is done by the ventricles, the bottom two of the heart's four chambers. The ventricles were what Jarvik and Kolff's artificial heart would replace.

Since this artificial heart was meant to be left permanently in the body, it had to be made of materials that were nontoxic and would not trigger an attack by the ever-vigilant immune system. They also had to be smooth enough that blood would not form clots on them. Such clots can block blood vessels. If the blockage occurs in the brain, the result is a stroke, which can cause paralysis or death.

Jarvik and Kolff decided to make their heart mostly out of polyurethane plastic. Its two chambers were shaped from a stiff form of the plastic. Each chamber had an inlet and an outlet valve on top, a flexible sheet or diaphragm stretched across the bottom, and an aluminum base. A tube led from the bottom of each chamber to the machine that provided power for the heart.

CONNECTIONS

HEARTS FROM SPACE

Critics have sometimes complained that the U.S. defense and space programs waste money that could be better spent helping people. In fact, research done for these programs has led to improvements in many fields of technology, including bionics. For instance, a metal called titanium was first used in missiles and submarines because it was lightweight, yet very strong. Today it is used in artificial hip joints for the same reason. Other materials developed for military or space use that are now used in artificial body parts include plastics such as Lucite, polyester meshes, lightweight fibers made of carbon, and self-sticking Velcro.

Now the designers had to find a good power source for their heart. After considering several possibilities, Kolff and Jarvik chose compressed air. The air came from a bulky machine called the heart driver, which was attached to the heart by 6-foot (1.8 m) tubes. Pulses of air under pressure pushed against the diaphragms at the bottom of the heart's hollow chambers, forcing blood out through the upper valves. From there the blood flowed into the body's two main arteries, the aorta and the pulmonary (lung) artery. In between pulses, the diaphragms relaxed, just as the heart muscle does. This let the artificial heart fill up with blood, which flowed in from veins attached to the heart's intake valves.

While Jarvik and Kolff were perfecting the design of the artificial heart, William de Vries, the surgeon on the Utah team, was learning the techniques needed to implant it by sewing artificial hearts into animals and human bodies. By the time de Vries put an artificial heart into a living human, he had already done the job on 200 calves and sheep and 20 bodies.

Barney Clark and "Bull" Schroeder

After many tests on animals, the Utah group was ready to try their heart in a human being. The model they planned to use was their seventh, so they called it the Jarvik-7. The FDA, which must approve all new medical treatments, gave its permission for the operation in 1981. It and the University of Utah, however, set up strict requirements for the sort of patient in whom the device was to be used. Finding someone who met all the requirements took over a year. This "perfect patient" was Barney Clark, a 61-year-old retired dentist from Seattle. Clark had a disease called cardiomyopathy, in which the heart swells and grows weak.

Clark and his wife, Una, listened while de Vries and Jarvik explained how they hoped the Jarvik-7 heart would help him and also the problems it might cause. Clark agreed to have the operation, not only in the hope that it would prolong his own life, but because he knew it would give the doctors information that would help them improve the heart.

I WAS THERE

A HAPPY MOMENT

Una Clark later described her feelings when she first saw her husband after he came out of the operating room with his new artificial heart.

> My knees buckled a little bit when I first saw Barney . . . with all the machinery and tubes, but instead of being grayish [as it had been], his skin was pink, and he looked wonderful. He began making all these motions, trying to tell us that he loved us. That was a very joyful time.

On December 2, 1982, in an operation that lasted 7½ hours, William de Vries removed the two ventricles of Barney Clark's failing heart. While a heart-lung machine kept Clark alive, de Vries attached the Jarvik-7 to the atria (upper chambers) of Clark's heart and to his body's two major arteries. One of the mechanical ventricles had to be replaced during the operation. Finally, however, the artificial heart settled down to beat reliably.

Barney Clark lived for 112 days with his new heart. During all that time he remained attached to the 375-pound (169-kg) air compressor, although he was able to walk a little in the hospital with the compressor following him on a cart. He finally died because his lungs, kidneys, and other vital organs, which had been

ISSUES

WAS IT WORTH IT?

In view of the suffering Barney Clark and William Schroeder underwent during the brief time they had their artificial hearts, critics questioned whether these difficult, expensive operations were worth doing. The wives of both men said after their deaths, however, that they believed their husbands would have gone ahead with the procedure even if they had known exactly what lay ahead. Una Clark told a reporter:

> Nothing worthwhile ever comes easily. I know that the artificial heart recipients have not received the quality of life they would have liked. But it will take many people to accomplish the goals. . . . If we succeed, it will have been worthwhile, and if we don't, at least we'll have shown the courage to fail. I know that my husband derived great satisfaction from his role in these experiments. I'm certain in my heart that he's very proud to have contributed.

damaged by his illness, failed. The artificial heart kept working to the end. A hospital technician had to shut it off.

During the next several years, 90 more people around the world received Jarvik-7 artificial hearts. Some additional patients were given a different model of artificial heart, the Penn State heart. This heart had been developed by bioengineers at Pennsylvania State University.

The person who lived the longest with an artificial heart was William J. ("Bull") Schroeder, a 53-year-old quality control checker at a military armaments plant. Like Clark, Schroeder had cardiomyopathy. De Vries implanted Schroeder's Jarvik-7 heart in 1986.

Schroeder survived for 620 days, or about 21 months, with his new heart. They were not very happy months, however. He suffered several strokes, caused by blood clots that formed in the heart and migrated to his brain. The strokes produced serious brain damage. Most of the other people who received artificial hearts also had strokes.

Because of these problems and the great expense of the treatment, many doctors and the public began to doubt the wisdom of using artificial hearts. In 1989, the FDA banned implantation of permanent artificial hearts in the United States. Artificial hearts are still used as a temporary "bridge to transplant," however.

New Artificial Hearts

Artificial hearts have improved greatly since Barney Clark's day. They are made of new plastics that are less likely to cause blood clots. Some experimenters are working on ways to coat heart surfaces with cells from the patient's own body. This would come closer to imitating the natural coating of blood vessels than any plastic can.

Some new artificial hearts, still being tested on animals at the Cleveland Clinic, differ from real hearts in a startling way: They don't beat. Scientists have found that the beat, or pulse, caused by

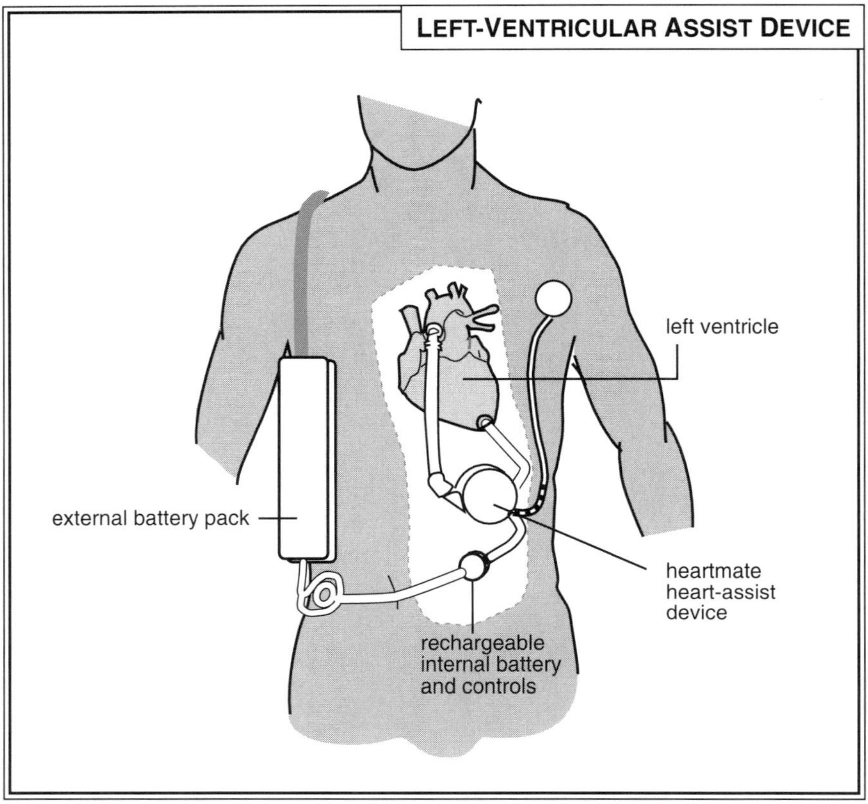

The Heartmate left-ventricular assist device (LVAD) is the closest thing to a permanent artificial heart in regular use today. Blood goes from the natural heart, which remains in place, to a pump in the abdomen that does most of the work of the heart's left ventricle. The pump is powered by a battery pack worn in a shoulder bag.

the expansion and contraction of the heart muscle is not really needed to keep blood flowing. The main job of the pulse in a natural heart is simply to give the heart muscle a chance to rest. Artificial materials do not need to rest. Because it does not constantly change size, an artificial heart that does not beat causes less damage to surrounding tissues. That reduces the chances of infection.

Meanwhile, Penn State bioengineers are experimenting with an artificial heart that can receive electricity through unbroken skin. Its batteries, carried in a shoulder pack, provide enough power to run the heart for about a day. The researchers hope to test their heart in humans by the year 2000. "Patients who have it could hold a job and climb stairs and do moderate exercise," says Penn State bionics expert Gerson Rosenberg.

About the closest thing to permanent artificial hearts on the market today are implanted machines called left-ventricular assist devices (LVADs). They do the pumping that would normally be done by the left ventricle, the heart chamber that works the hardest. Since 1986, LVADs have been implanted in more than 700 people. The devices kept the people alive for up to 17 months while they waited for transplants.

One type of LVAD is called the Heartmate. It is implanted in the abdomen and connected by a short tube to the person's natural heart, which remains in place. Blood flows through the tube from the left ventricle to the LVAD. When the Heartmate's pump is filled with blood, sensors turn on its electric motor. Power for the motor comes from a battery pack that, like the one for the new Penn State heart, is worn in a shoulder bag. The motor pushes a pistonlike plate that, in turn, pushes the blood into the aorta.

Challenges to Artificial Organs

Artificial kidneys and hearts are fairly easy to design. Organs such as livers and lungs are much harder because their structure and functions are so complex. A liver, for instance, is a whole chemical factory. It removes some substances from the blood, adds others, and changes still more into different substances. Because of this complexity, artificial livers and lungs probably will not exist for many years.

All present and future artificial organs face two key problems. One is infection. Despite the great strides in antisepsis and asepsis made since Joseph Lister's time, any break in the skin (such as is required, for instance, to connect blood vessels to a dialysis machine or an artificial heart to a power source) still can let dangerous microbes invade the body. Antibiotics can control many of these microorganisms, but some are resistant to drugs.

Infections have threatened or ended the lives of many people with artificial organs. The other problem is the body's reaction to foreign materials placed in it. This reaction may take the form of blood clots, attempts to dissolve the foreign material, or attacks by the immune system.

Scientists will go on trying to solve these problems. Transplanted natural organs may work better than artificial ones, but, as the next chapter will show, the number of people who need transplants is far greater than the number of organs available. In addition, thousands of people are considered too old, too ill, or otherwise unqualified to receive transplants. Permanent artificial organs may be these people's only hope.

Chronology of Artificial Organs

February 14, 1911	Willem Kolff is born in Leyden, Holland
1913	Artificial kidney used in dogs
1938	Kolff begins work on artificial kidney
March 17, 1943	Artificial kidney first used on a human being
September 1945	First life saved by artificial kidney
May 11, 1946	Howard Jarvik is born in Midland, Michigan
1950	Kolff emigrates to United States
1967	Moves to University of Utah
1971	Jarvik begins work on artificial heart
December 2, 1982	First artificial heart implanted in human being (Barney Clark)
1983	Clark dies after living 112 days with artificial heart

1986 Artificial heart implanted in William Schroeder, who survives for 21 months

1989 Implantation of permanent artificial hearts banned in United States

Further Reading

Berger, Melvin. *The Artificial Heart.* New York: Franklin Watts, 1987. For young adults. Describes the development of artificial hearts and how they work.

Facklam, Margery and Howard. *Spare Parts for People.* San Diego, Calif.: Harcourt Brace Jovanovich, 1987. For young people. Includes chapters on artificial hearts and artificial kidneys.

"Jarvik, Robert K(offler)." *Current Biography 1985.* New York: H. W. Wilson, 1985.

"Kolff, Willem Johan." *Current Biography 1983.* New York: H. W. Wilson, 1983.

Metos, Thomas H. *Artificial Humans.* New York: Julian Messner, 1985. For young people. Includes material on artificial hearts and kidneys.

Robinson, Donald. *The Miracle Finders.* New York: David McKay, 1976. Includes dramatic account of Willem Kolff's invention of the artificial kidney.

NOTES

p. 76 "I saw them . . ." Quoted in Donald Robinson, *The Miracle Finders* (New York: David McKay, 1976), p. 71.

p. 77 "the father . . ." University of Utah handout, 1997.

p. 77 "It just shows . . ." Quoted in Janice M. Cauwels, *The Body Shop* (St. Louis: C. V. Mosby, 1986), p. 129.

p. 80 "I take enthusiastic . . ." Quoted in Cauwels, p. 13.

p. 82 "I decided . . ." Quoted in Melvin Berger, *The Artificial Heart* (New York: Franklin Watts, 1987), p. 47.

p. 85 "My knees . . ." Quoted in Cauwels, p. 184.

p. 86 "Nothing worthwhile . . ." Quoted in Cauwels, p. 192.

p. 89 "Patients who . . ." Quoted in Marsha F. Goldsmith, "After Two Decades, Penn State Researchers May Be Near Perfecting Replacement Hearts and Lungs," *Journal of the American Medical Association*, February 24, 1993, p. 968.

6

New Hearts for Old

NORMAN SHUMWAY, CHRISTIAAN BARNARD, AND ORGAN TRANSPLANTS

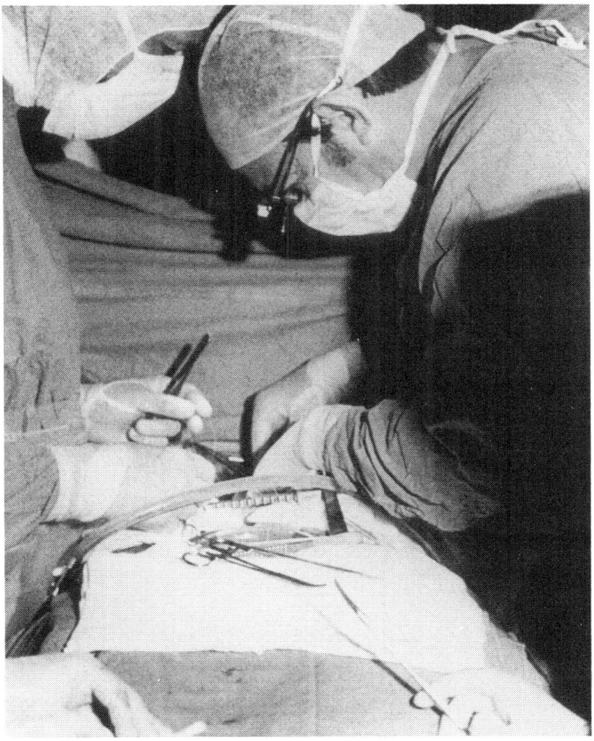

South African surgeon Christiaan Barnard performed the first human heart transplant at Groote Schuur Hospital in Cape Town. (Courtesy University of Cape Town, Cape Town, South Africa)

Time called it "the surgical equivalent of [climbing] Mount Everest." *Newsweek*'s story about it was headed "Miracle in Cape Town" and proclaimed that it was "perhaps the most remarkable bit of surgical virtuosity in the history of medicine." It made handsome young South African surgeon Christiaan Neethling Barnard and his patients, Louis Washkansky and Denise Darvall, famous around the world.

Like the implantation of an artificial heart into Barney Clark almost exactly 15 years later, it could be considered both a stunning success and an almost complete failure.

It was the first transplantation of a living heart from one human being to another.

Christiaan Barnard's transfer of a heart from 25-year-old Darvall, killed in an auto accident, to 55-year-old Washkansky on December 3, 1967, was not the beginning of the story of organ transplants, however. Like the first artificial organs, that had occurred many years before.

Mysterious Failures

Legends tell of a Chinese sage and a pair of saintly European brothers who transferred organs from one person to another without harm to their patients. In real life, though, any surgeon who dared to try an organ transplant was sure to fail. A transfusion of blood might work now and then, but not the transfer of other tissues or organs.

It was not that the surgery was so difficult. Near the beginning of the 20th century, two surgeons working in the United States, Alexis Carrel and Charles Guthrie, solved most of the problems that would occur in the operating room during an organ transplant. They transferred kidneys, hearts, and even heads from one dog to another. After a day or two at most, though, the donor organs shriveled and died. Something in the recipient's body destroyed them, but at first no one knew what, let alone how or why.

Karl Landsteiner's work with blood types offered the first clue by showing that the failure of some blood transfusions—which were transplants of a sort—was due to the immune system. Something besides blood types had to be involved in the failure of organ transplants, though, because such failures occurred even when donors' and recipients' blood types were the same.

A British scientist named Peter Medawar finally solved the mystery in 1944. He showed that cells in the body's solid tissues, like those in the blood, had antigen markers on their surfaces. The immune system attacked cells carrying tissue antigens different from a person's own, just as it did with blood cell antigens. Whereas many people had the same major blood antigens, however, only identical twins, who have exactly the same genes, had all the same tissue antigens. Medawar's work suggested that organ transplants, except between identical twins, might remain an impossible dream.

Transplanted Kidneys

Fortunately, anyone who made such a dark prediction would have been wrong. The first successful transplant of internal organs took place 23 years before Christiaan Barnard picked up his scalpel. Like the first artificial organ, the first transplanted organ was a kidney. As Medawar's research would have predicted, the donor and the recipient of the transplant were identical twins. (A person can remain healthy with only one kidney, so one twin chose to give up his second kidney to save his dying brother.) Surgeon Joseph Murray carried out the transplant operation at Peter Bent Brigham Hospital in Boston in 1954.

Now came the hard part. In order for transplants between people other than identical twins to succeed, doctors had to find ways to keep the recipient's immune system from destroying the donor organ. Researchers had learned as early as 1951 that a hormone (a type of chemical that carries messages from one part of the body to another) called cortisone suppressed the immune system enough to help transplanted skin survive, at least temporarily.

By itself, however, cortisone's effect was not great enough to permit successful organ transplants.

Doctors tried other methods so drastic that they destroyed the immune system almost completely. Some bombarded transplant recipients' bodies with high-energy radiation. Others gave their patients 6-mercaptopurine, a drug used to treat blood cell cancers because it destroyed immune system cells. A combination of radiation and 6-mercaptopurine was used in the first successful kidney transplant between nonidentical (fraternal) twins, which took place in 1959.

The trouble with destroying the immune system was that it left the body defenseless. Like people whose immune systems have been ruined by AIDS, most early transplant recipients suffered one microbe-caused illness after another. They also were likely to develop cancer because of cell damage from the treatments they took. Doctors therefore looked for ways to weaken the immune system that were less hard on the body.

In 1962, University of Colorado researcher Thomas Starzl and his coworkers began combining a newly discovered drug called

KEY CONCEPTS

brain death a standard that defines death as the point at which the brain no longer functions

cyclosporine a drug that prevents transplant rejection without destroying the immune system

steroids hormones (message-carrying body chemicals) related to cortisone

tissue engineering a new medical field whose scientists grow body tissues on artificial frameworks azathioprine, a milder relative of 6-mercaptopurine, with steroids—hormones related to cortisone. This new "cocktail" suppressed the immune system enough to permit transplants to survive, but it did not leave the body totally vulnerable to disease.

Using this treatment, Joseph Murray and a few other surgeons made successful kidney transplants between unrelated people for the first time in 1962. Like most organs transplanted later, the kidneys in these operations came from the bodies of relatively young, healthy people who were killed suddenly by such causes as car accidents or strokes. Recipients' tissue antigens were matched as closely as possible to those of donors, because the more antigens the two had in common, the less the immune system had to be suppressed. By the time of Christiaan Barnard's heart surgery, some 1,500 kidneys had been transplanted worldwide.

"The Way Is Clear"

If kidneys could be transplanted, some surgeons felt, there was no reason why hearts couldn't, too. Chief among these believers was a California surgeon named Norman Edward Shumway. Born in Kalamazoo, Michigan, on February 9, 1923, Shumway joined the Stanford School of Medicine in 1958. There he and a younger surgeon, Richard Lower, experimented with dogs to develop new techniques in heart surgery.

Among other things, Shumway's animal experiments taught him that a heart transplant was most likely to succeed if the tops of the two atria (upper chambers) of the recipient's heart were left in place. Because the atria are connected to several of the body's major blood vessels, leaving the atria intact greatly reduced the number of blood vessels a surgeon had to rejoin. It cut operating time in half.

In 1967, Shumway wrote in a medical journal that, from a surgical point of view, "the way is clear for heart transplant." Any one of a number of surgeons could carry out the operation. The first one to do so would be the one who found a compatible

Stanford University surgeon Norman Shumway developed many of the techniques used in heart transplant operations. He continued to carry out heart transplants even when other surgeons lost faith in them. (Courtesy Stanford University News Service)

recipient and donor first. The lucky surgeon turned out to be Christiaan Barnard.

Barnard had already shown that he seldom let opportunities slip past him. Born in 1923 to a minister's family in Beauford West, a small settlement in South Africa, he had grown up in poverty. His parents still somehow saved enough money to send him and his three brothers to college. (His mother always told him, "I expect you to be first. Not second or third, but first.") Barnard studied surgery at the University of Minnesota, where he was a classmate of Norman Shumway's. He later learned the new heart surgery techniques that Shumway developed. At the time Barnard made medical history, he was head of cardio-thoracic (heart and chest) surgery at Groote Schuur Hospital, a large teaching hospital that is part of the University of Cape Town.

Unlike Barnard, Louis Washkansky wasn't looking for a place in history, but he was willing to accept one. The 55-year-old grocer, born in Lithuania, had suffered several massive heart attacks. By late 1967, he knew he would die soon unless some new treatment was found. When Barnard asked him whether he would agree to a heart transplant if a suitable donor could be found, Barnard said he would come back for Washkansky's answer in two days. Washkansky, however, decided in two minutes. "Go ahead," he said.

Denise Darvall had no wish at all to enter history. She just wanted to pick up a cake to take to friends during a holiday visit. On the afternoon of December 2, Darvall, a bank clerk, and her mother, Myrtle, walked to a bakery to get the cake. Edward Darvall, Denise's father, waited in their car nearby. As the two women came back across the street with their purchase, they were hit by a speeding car. Myrtle Darvall was killed instantly, and Denise's hips and skull were crushed. She was barely alive when she was brought to Groote Schuur Hospital. There was no hope of saving her.

A Historic Operation

Christiaan Barnard heard about Denise Darvall's arrival and wondered if she might be the heart donor he was looking for. Nervously, he approached the grieving Edward Darvall and described Washkansky's situation to him. "We have done our best,

SOCIAL IMPACT

A NEW DEFINITION OF DEATH

Organ transplants changed the way society defined death. The change did not happen overnight, though. A year after the first successful heart transplant operation, the American Medical Association redefined death as brain death: the point when "the individual shows no responsiveness and awareness of any kind, no movement or spontaneous breathing and no reflex [automatic nervous system] activity." Laws defining death as "when the heart stops beating" remained in force, however.

This conflict of definitions led to some strange courtroom scenes. For instance, in a 1973 trial, a defense lawyer in California tried to save his client by accusing Norman Shumway of murder! The lawyer claimed that the death of the man whom the defendant was accused of killing did not occur when the man was shot, rendering him brain dead, but rather when Shumway removed the man's heart for a transplant. Fortunately for Shumway, the jury did not accept that argument.

and there is nothing more we can do to help your daughter," Barnard said. "You can do us and humanity a great favor if you will let us transplant your daughter's heart."

Fortunately, Darvall was able to rise above his sorrow. "If there's no hope for her, then try to save this man's life," he replied.

Now it was just a matter of waiting for Denise Darvall to die. The usual definition of death was "when the heart stops beating." Unfortunately, body tissues begin breaking down very quickly after this point. Looking ahead to a possible future era of transplants, some doctors had suggested changing the definition of death to "the point at which the brain no longer functions." Because the heart beats automatically, without direct input from the brain, it can continue for a while after a person is "brain dead."

WASHKANSKY HEART TRANSPLANT

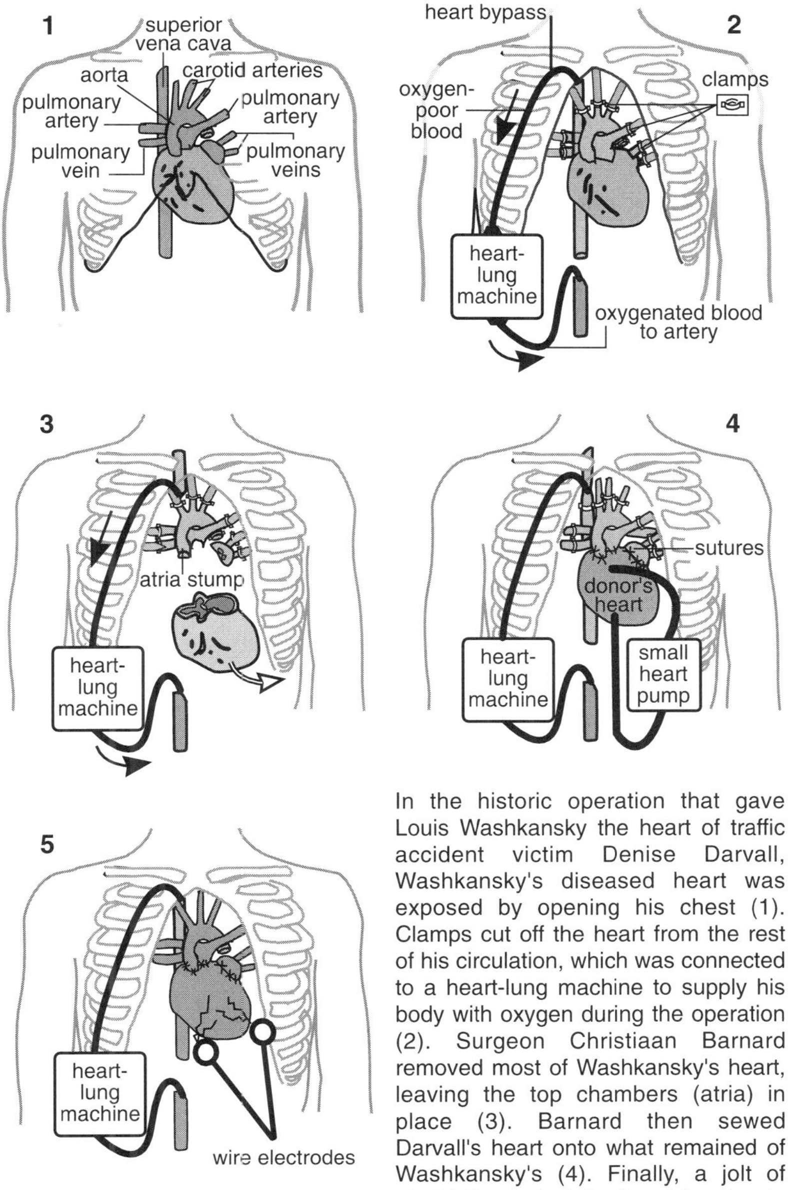

In the historic operation that gave Louis Washkansky the heart of traffic accident victim Denise Darvall, Washkansky's diseased heart was exposed by opening his chest (1). Clamps cut off the heart from the rest of his circulation, which was connected to a heart-lung machine to supply his body with oxygen during the operation (2). Surgeon Christiaan Barnard removed most of Washkansky's heart, leaving the top chambers (atria) in place (3). Barnard then sewed Darvall's heart onto what remained of Washkansky's (4). Finally, a jolt of electricity started the heart beating in its new location (5).

Barnard might have preferred that the brain definition of death be applied to Denise Darvall. Groote Schuur, however, used the more common heart definition. Thus, the brain-dead Darvall lay in one operating room, her breathing maintained by a respirator, and Louis Washkansky lay in another while Barnard and his surgical team waited for Darvall's heart to stop.

Denise Darvall's heart finally stopped beating at 2:15 A.M. on December 3. A member of Barnard's team opened Darvall's chest, exposed the heart, and connected it to a heart-lung machine. The machine circulated blood through Darvall's heart, helping to preserve it.

Meanwhile, in the operating room next door, the anesthesiologist quickly put Louis Washkansky to sleep. Barnard opened Washkansky's chest and attached his major blood vessels to a second heart-lung machine. He then removed first Darvall's heart and then Washkansky's. Following Shumway's technique, he left the top of Washkansky's heart in place. He carefully sewed Darvall's heart into the spot where Washkansky's had been.

I WAS THERE

HOLDING A HEART

Christiaan Barnard later described what he felt was the most dramatic moment in his operation on Louis Washkansky:

> My moment of truth—the moment when the enormity of it all really hit me—was just after I had taken out Washkansky's heart. I looked down and saw this empty space. . . . The realization that there was a man lying in front of me without a heart but still alive was, I think, the most awe-inspiring moment of all.

Finally, Barnard was ready for the climactic moment: taking off the clamps that had closed the blood vessels near what was left of Washkansky's heart and finding out whether Darvall's heart would function in his chest. The surgeons gave the transplanted heart a jolt of electricity to start it up. The heart leaped and then settled into a steady rhythm. "Christ, it's going to work!" Barnard burst out. When he was sure that the new heart was doing its job, he sewed up Washkansky's chest. At the end of 4¾ hours of grueling work, he then left the operating room. "I need a cup of tea," he said.

Four days after the surgery, Washkansky was joking with Barnard and news reporters. "I'm a Frankenstein [monster] now," he chuckled. "I've got somebody else's heart."

Boom, Bust, Boom

Barnard's triumph was spectacular but short-lived. Washkansky was treated with a combination of azathioprine, steroids, and radiation to keep his immune system from rejecting Darvall's heart. This onslaught on his immune system left him open to infection, and he developed pneumonia. He died of it just 18 days after his operation.

Washkansky's death cast little pall over the excitement that the operation produced. Suddenly, every major surgeon and hospital seemingly wanted to do a heart transplant—whether it helped their patients or not. Some transplant recipients lived for months, but others died within hours.

Christiaan Barnard, lucky again, claimed the longest survivor during this early period of heart transplants. That patient was Philip Blaiberg, a 58-year-old retired dentist. Blaiberg lived for 593 days—over a year and a half—with his new heart.

In 1968, 105 heart transplant operations were performed in different parts of the world. Then, however, as would happen later with artificial hearts, the excitement died down and disillusion set in. Of the 150 people who received new hearts during 1968 and

PARALLELS

HEART TRANSPLANTS AND ARTIFICIAL HEARTS

The technologies of organ transplants and artificial organs developed in similar ways. Both started with kidneys but attracted widespread attention only when hearts began to be replaced. The popularity of heart transplants and of artificial hearts went through similar "boom and bust" cycles 15 years apart. Thanks to a medical breakthrough, however, heart transplants have again become common. This has yet to happen with implantation of artificial hearts.

1969, only about 30 were still alive at the end of 1969—an unimpressive 20 percent survival rate. The small gain in lifespan hardly seemed worth the operation's considerable expense.

By the early 1970s, heart transplantation had all but stopped. Only the determined Norman Shumway refused to give up on the procedure. Shumway, a quiet man, once said that Christiaan Barnard, who had clearly enjoyed the superstar role of being the first heart transplant surgeon, "did us a favor by performing the operation first. He drained off a lot of the sensational publicity and made it easier for us to keep the heart transplant in context here [at Stanford]." During the 1970s, Shumway steadily improved his surgical technique and use of antirejection drugs. As a result, his patients survived longer and longer. By 1980, they were living almost as long as the average kidney transplant patient.

Transplants finally became truly practical, though, because of a bit of dirt that a collector picked up on a trip to Norway in 1970. The dirt made its way to the Swiss drug company Sandoz, whose scientists discovered that a fungus in the sample made a substance they had never seen before. They named this chemical cyclosporine.

Tests revealed that cyclosporine kept the immune system from reacting to foreign tissues, yet it suppressed the system's overall activity much less than the azathioprine-plus-steroid treatment.

Cyclosporine began being given to heart transplant patients in the United States in 1980 and to recipients of other organ transplants in 1983. Combined with steroids, it brought about a revolution. Pioneer transplant researcher Thomas Starzl called cyclosporine "the key that unlocks the door to transplants." For example, cyclosporine boosted the percentage of Norman Shumway's heart transplant patients who lived at least a year from 63 percent to 83 percent.

Thanks to cyclosporine and other new drugs, transplants of kidneys, hearts, and livers have now become almost routine. Transplants of lungs or heart-and-lung combinations are more difficult, but they, too, have been done successfully. Thousands of people have survived for 20 years or more with transplanted organs. Many lead active, even athletic, lives.

Desperate Shortages

In spite of these advances, important problems still haunt transplant surgery. Scientists would like to find better ways to control rejection and to preserve donated organs, for instance. A donated kidney can survive up to 72 hours, but a heart, liver, or lungs usually must be transplanted within 4 to 6 hours of the donor's death.

By far the greatest problem with transplants, though, is simply that far more people need organs than donate them. As transplant operations have become more successful, the demand for organs has grown greatly. Unfortunately, the supply of donor organs has remained about the same as it was in the 1980s.

Polls have shown that 85 percent of Americans think that organ donation is a good idea—yet many never sign forms giving permission to donate their organs or tell family members about their wishes. (Even if a person has signed an organ donor card, his or

TRENDS

DONOR ORGANS: SUPPLY AND DEMAND

The following rounded figures show the painfully increasing gap between the number of organs transplanted and the number of people on the waiting list for transplants.

Year	Number on Waiting List	Transplants
1991	25,000	17,000
1992	29,000	17,000
1993	32,000	18,000
1994	35,000	19,000
1995	44,000	20,000

her next of kin must still give permission for the donation, so talking to family members is very important.) Estimates suggest that 14,000 potential organ donors die in the United States each year, yet fewer than 4,500 actually become donors.

Because the supply of organs is so limited, people who need transplants are placed on waiting lists. Organ donor networks such as the United Network for Organ Sharing, which controls the distribution of transplanted organs in the United States, keep the lists. The people must wait, sometimes for years, for a suitable organ to become available. About a third of would-be transplant recipients die before their names come up.

Easing the Transplant Crunch

Several solutions have been proposed for the transplant dilemma. Some are social or legal, such as requiring doctors and nurses to ask all people entering a hospital, or their families if they cannot communicate, to state their wishes about organ donation. Other solutions are technical, such as building artificial organs.

A more radical solution would be to transplant animal organs into humans. Naturally, organs from a different species contain far more different antigens than do organs from any member of the same species. Preventing rejection of animal organs thus would be much harder than preventing rejection of human ones. Researchers believe that it may eventually be possible, however.

If animal organs can be used, it might seem logical to take them from humans' closest relatives: chimpanzees, baboons, or other primates. However, some primates, including chimpanzees, the primates to which humans are most closely related, are endangered. Even more important from humans' point of view, viruses living in a primate organ donor might be transferred to a human recipient along with the organ. The virus that causes AIDS is thought to have come from monkeys and spread somehow to

I WAS THERE

THE WAITING IS THE WORST PART

Pat DeFries, who received a heart transplant in the 1980s, described the frustration of waiting for a donor organ:

> The waiting is the worst part. First you go through all the red tape while the hospital tries to find the funding to pay for the operation. . . . The emotional ups and downs are unreal. You find yourself waiting for weekends and holidays because there are more car accidents then. I know it sounds terrible, but all you can think of is that your heart is dying, and someone else has to die for you to live.

ISSUES

WHO SHOULD GET TRANSPLANTS?

The shortage of donor organs forces hard decisions. In the past, networks usually gave available organs to the sickest people because their need was greatest. In 1996, however, the United Network for Organ Sharing reversed this policy for livers, deciding instead to give them to people who have been ill only a short time because such people have less damage to the rest of their bodies than do people who have been sick a long time. They therefore have a greater chance of being helped by the transplants. Similar changes may be made for other organs.

humans. Some other primate virus, passed to people through organ donation, might cause an even worse plague.

If organ transplants from animals ever do become a reality, the organs will probably come not from primates but from pigs. Pigs are cheap to raise, and they mature quickly. They have more genes in common with humans than people might like to think, yet they are different enough that viruses living in them are not too likely to infect humans. Genetic engineers have already bred pigs containing genes that produce human antigens. The tissues of these pigs thus should appear human to people's immune systems and might work for transplants.

In the long run, the best way to replace a diseased organ may be to grow a new one. Scientists in the new field of tissue engineering can already grow skin, bone, and some other tissues in the laboratory and use them in transplants. Tissue cells are usually "seeded" onto a netlike framework made of synthetic or natural material. This material contains no antigens and can dissolve in the body. Tissue and framework are transplanted together. In time the framework is absorbed, leaving only normal tissue behind.

Growing organs, which have a complex structure and usually contain several kinds of tissue, will be much harder than growing tissues. In the future, however, tissue engineers may be able to use new discoveries in genetics and molecular biology to shape whole hearts, livers, or kidneys. They may even make them from the recipient's own cells, which would avoid the rejection problem. "If we can't find replacement organs," says Joseph Vacanti, a leader in the tissue engineering field, "we'll build them."

Chronology of Organ Transplantation

early 1900s	Alexis Carrel and Charles Guthrie transplant organs in animals
November 8, 1922	Christiaan Barnard is born
February 9, 1923	Norman Shumway is born
1944	Peter Medawar shows that tissues contain antigens different from blood group antigens
1954	First successful kidney transplant between identical twins
1961	Norman Shumway begins developing technique for transplanting hearts
1962	Azathioprine used in first successful kidney transplant between unrelated people
December 3, 1967	First human heart transplant done by Christiaan Barnard in Cape Town, South Africa
1968	First heart transplant recipient dies
1983	Cyclosporine approved for transplant patients

Further Reading

Barnard, Christiaan. *One Life*. New York: Macmillan, 1969. Barnard's autobiography, including a detailed account of the first human heart transplant.

Dreher, Nancy. "How One Person Is Living with a Serious Disease." *Current Health 2*, March 1996. Interview with Jerry Cound, a young man who underwent two heart transplants as a teenager; describes his experiences vividly.

Durrett, Deanne. *Organ Transplants*. San Diego, Calif.: Lucent Books, 1993. Overview of the subject for young adults.

Hoffman, William, and Jerry Shields. *Doctors on the New Frontier*. New York: Macmillan, 1981. Devotes a chapter to Norman Shumway.

Lee, Sally. *Donor Banks*. New York: Franklin Watts, 1988. For young adults. Has chapters on organ and tissue banks.

Leinwand, Gerald. *Transplants: Today's Medical Miracles*. New York: Franklin Watts, 1992. For young adults.

United Network for Organ Sharing. www.ew3.attnet/unos. Includes news bulletins, statistics, reports, and resources related to organ transplantation.

NOTES

p. 94 "the surgical equivalent . . ." "The Ultimate Operation," *Time*, December 15, 1967, p. 64.

p. 94 "perhaps the most . . ." "The Heart: Miracle in Cape Town," *Newsweek*, December 18, 1967, p. 86.

p. 97 "the way is clear . . ." Quoted in Roger Rapoport, *The Superdoctors* (Chicago: Playboy Press, 1975), p. 130.

p. 99 "I expect . . ." Quoted in Donald Robinson, *The Miracle Finders* (New York: David McKay, 1976), p. 203.

p. 99 "Go ahead" Quoted in "The Ultimate Operation," p. 64.

pp. 99–100 "We have done . . ." "If there's no hope . . ." Quoted in William Hoffman and Jerry Shields, *Doctors on the New Frontier* (New York: Macmillan, 1981), p. 15.

p. 100 "the individual shows . . ." Quoted in Hoffman and Shields, p. 34.

p. 102 "My moment . . ." Quoted in Peter Hawthorne, *The Transplanted Heart* (Chicago: Rand McNally, 1968), p. iii.

p. 103 "Christ, it's . . ." "I need a cup . . ." Quoted in "The Ultimate Operation," p. 65.

p. 103 "I'm a Frankenstein . . ." Quoted in Hoffman and Shields, p. 17.

p. 104 "did us a favor . . ." Quoted in Rapoport, pp. 131–32.

p. 105 "the key . . ." Quoted in Melvin Berger, *The Artificial Heart* (New York: Franklin Watts, 1987), p. 31.

p. 107 "The waiting is . . ." Quoted in Margery and Howard Facklam, *Spare Parts for People* (San Diego, Calif.: Harcourt Brace Jovanovich, 1987), pp. 33–34.

p. 109 "If we can't . . ." Quoted in Susan Reed, "Toward Remedying the Organ Shortage," *Technology Review*, January 1994, p. 45.

7

The Better to See You With

GODFREY HOUNSFIELD, RAYMOND DAMADIAN, AND MEDICAL IMAGING

British engineer Godfrey Hounsfield won a Nobel Prize in 1979 for his invention of computerized axial tomography (CAT or CT), one of several techniques developed in recent years that greatly improve doctors' ability to see inside the body. (Courtesy Albert and Mary Lasker Foundation and National Library of Medicine)

In the early days of the television series *Star Trek*, the spaceship's doctor found out what was wrong with a person simply by waving a handheld device in front of the person's body. Doctors don't yet have anything as portable as Dr. McCoy's high-tech magic—but, thanks to several inventors, they have more bulky tools that do almost the same job. With these machines they can look inside the body without surgery and in some cases without radiation. The machines create images that Wilhelm Röntgen and his early radiologist successors could hardly dream of.

Doctors use these new imaging tools to detect disease earlier and to design treatment more precisely than has ever been possible before. The result is the rescue of innumerable lives. Over and over, while science writer Howard Sochurek was researching a story for *National Geographic* on the new medical imaging, doctors told him, "I couldn't have saved this patient 10 years ago."

How Is a Baby Like a Submarine?

Ultrasound, the oldest of the new imaging techniques, grew out of a wartime invention called sonar. Sonar spotted submarines under water by sending very high-pitched sound waves through the water and analyzing the echoes that bounced back. The timing and pattern of the echoes made images that showed the location of solid objects in the water.

In the early 1950s, a Scottish doctor named Ian Donald realized that, strange as it seemed, an unborn baby (a fetus) and a submarine had something in common. Both were solid objects floating in an "ocean" of liquid. Donald wondered whether something like a sonar detector could form a picture of a fetus in a woman's womb. If it could, such a technique might be used on pregnant women instead of dangerous X-ray examinations.

Donald and others built the first ultrasound machines from leftover sonar equipment. The machines began to be used in medicine in the late 1950s. Today, ultrasound examines not only babies but abdominal organs, blood flow, and tumors.

Because it uses no radiation, ultrasound is the only kind of medical imaging recommended for pregnant women. Ultrasound tests are relatively inexpensive and simple to give. A technician just smears a jellylike substance on the part of the patient's body to be examined. A handheld device that sends out and receives

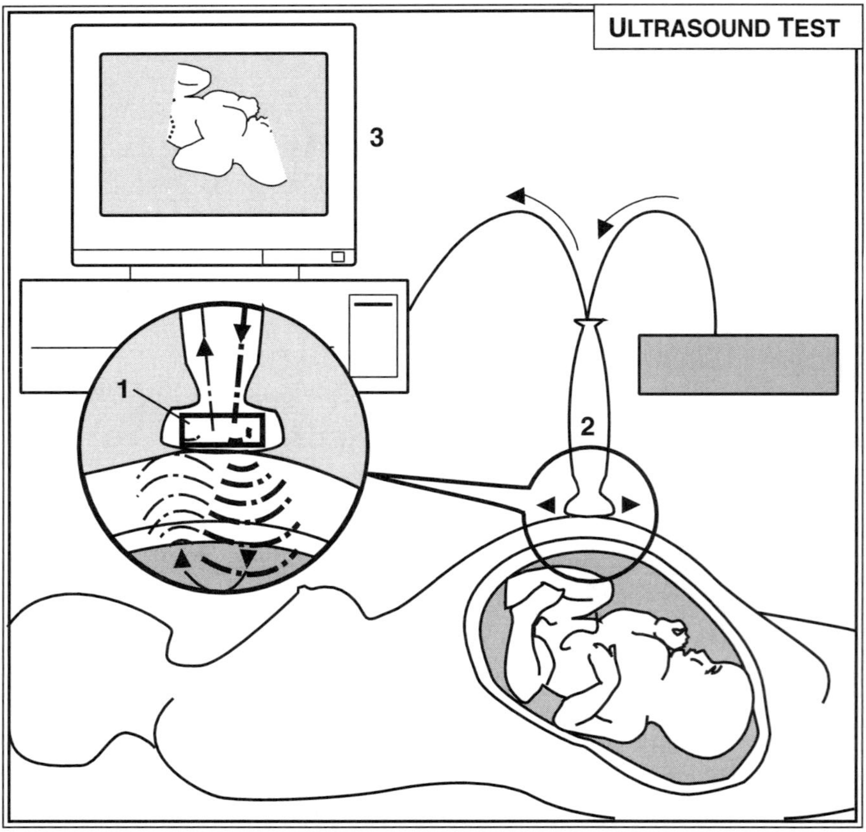

In an ultrasound test to check the health of a fetus inside a pregnant woman, a jellylike substance is rubbed on the skin of the woman's abdomen. A technician then runs a handheld device (2) over the skin. A crystal in the device (1) translates electric signals into sound waves too high-pitched to hear. These waves penetrate the woman's body and bounce off the fetus, producing echoes. The device detects these echoes and retranslates them into electric signals. A computer "reads" these signals and transforms them into an image (3).

KEY CONCEPTS

CAT or CT (computerized axial tomography) a process in which a computer combines multiple X-ray pictures to show a "slice" through the body

NMR or MRI (nuclear magnetic resonance—later magnetic resonance imaging) a process in which atoms' reactions to a magnetic field and radio waves are used to form images of body tissues

ultrasound use of the pattern of echoes produced by high-frequency sound waves to produce images

sound waves is then slid over the skin. A computer translates the sound echoes, which it receives as electrical pulses, into an image that the technician sees on a monitor screen.

Reinventing the X Ray

From the start, X-ray photos had serious disadvantages as views into the body. First, the only things that showed up really clearly on them were bones. Doctors learned how to make some organs show in X-ray photographs, but such photos never revealed much detail. Second, X-ray pictures were two-dimensional. If an organ lay behind bones or other organs, doctors could not get a good look at it. Finally, X-ray pictures could not show motion, such as the beating of a heart.

All that began to change in the 1970s, thanks to a British electronics engineer named Godfrey Hounsfield. Born on August 28, 1919, in Newark, England, Hounsfield learned about imaging during World War II. Afterward he did computer research for a large British company called EMI. One day in 1967, while he was walking through the hills near his home, it occurred to him that if computers could recognize printed characters on a page, they might be able to interpret X-ray photographs, too. He also became interested in an imaging method called tomography, which shows thin cross-sections, or "slices," through three-dimensional objects.

Hounsfield thought he might be able to combine X rays, tomography, and computers. If an X-ray source and detectors were rotated around a body or body part, he reasoned, they could produce many images from different angles. A computer could combine these images into a single picture that showed a cross-section of the body. Such an image should be more accurate and detailed than a standard X-ray photo. Hounsfield's system became known as computerized axial tomography (CAT or CT).

Computers of the late 1960s were immense, but they were much less powerful than the computers of today. At first they could not give Hounsfield the results he wanted. As computers improved, however, so did the quality of his images.

Hounsfield tested the first models of his CT scanner by making pictures of preserved human brains and of the bodies and brains of freshly killed animals. He was pleased to see that his first photograph of a preserved brain clearly showed its two types of tissue, the gray and white matter—something no ordinary X-ray photograph could do.

Hounsfield's first scans took nine days apiece to make. Each produced 28,000 measurements from the X-ray detectors. A computer needed 2½ hours to process all this information and another 2 hours to turn it into an image. When he began using a stronger X-ray source, however, he reduced the scanning time to nine hours.

EMI began selling CT scanners in 1972. At first they could photograph only a patient's head. The first scan of a living person, made on October 4, 1972, helped doctors identify a brain tumor. Later scans revealed the location of blood clots in the brain.

CT scanners that could examine any part of the body became available in 1975. They showed clear differences between types of tissue and between normal tissues and cancers. Scanning time dropped to less than a minute.

The Glamour Machine

CT scanners were extremely expensive at first ($500,000 or more apiece), but the information they produced was so useful that they quickly became popular. As had happened with X rays, both doctors and ordinary citizens became excited about the new machines. For instance, the *Wall Street Journal* said on December 10, 1975, "It's revolutionary. This is the glamour machine." Some of the glamour rubbed off on Godfrey Hounsfield when he won the Nobel Prize in medicine in 1979.

Today there are more than 10,000 CT scanners in the United States alone. Most large and medium-sized hospitals have one. A

CONNECTIONS

IMAGE OF A MUMMY

CT and other new imaging technologies are used in many fields besides medicine. Archaeologists, for instance, recently gave a CT scan to a "patient" who had been dead for 500 years. She was a 14-year-old girl from Peru's Inca civilization. The girl had been killed as a religious sacrifice and left on top of a mountain, where her body dried into a mummy. The scan showed that she had been killed by a blow to the head, not by suffocation as the scientists had first thought. It also showed that she had been quite healthy when she died.

CT scan machine looks like a giant square doughnut with a table in the middle. The person being scanned lies on the table, which can be slid forward as little as $^{1}/_{100}$ of an inch at a time. A computer controls both table and scanner.

X-ray sources in the inner part of the doughnut send out a fan-shaped beam of rays. Patient and table are positioned so that the rays pass through the part of the patient's body that a doctor wants to study. The rays are picked up by a row of detectors on the side opposite the source. The detectors register signals of different strengths, depending on whether the rays pass straight

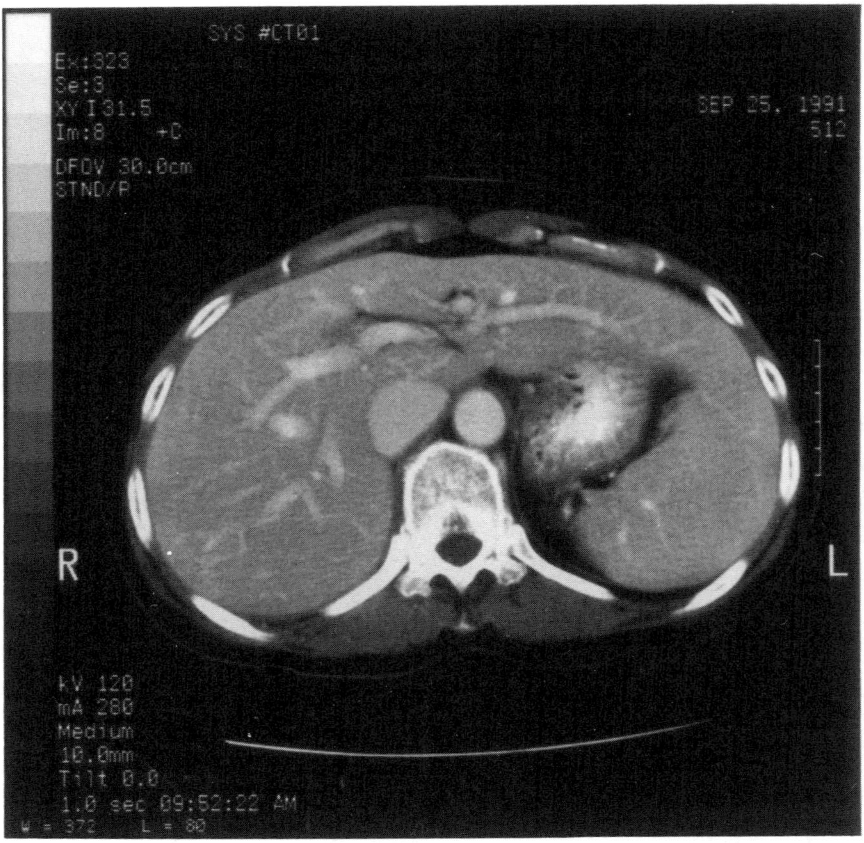

A CT scan uses X rays, but it can show much more than traditional X-ray photographs could. This scan shows a "slice" or cross-section through the middle of the body, revealing not only the backbone (center bottom) but the soft tissues of several organs. (Courtesy GE Medical Systems)

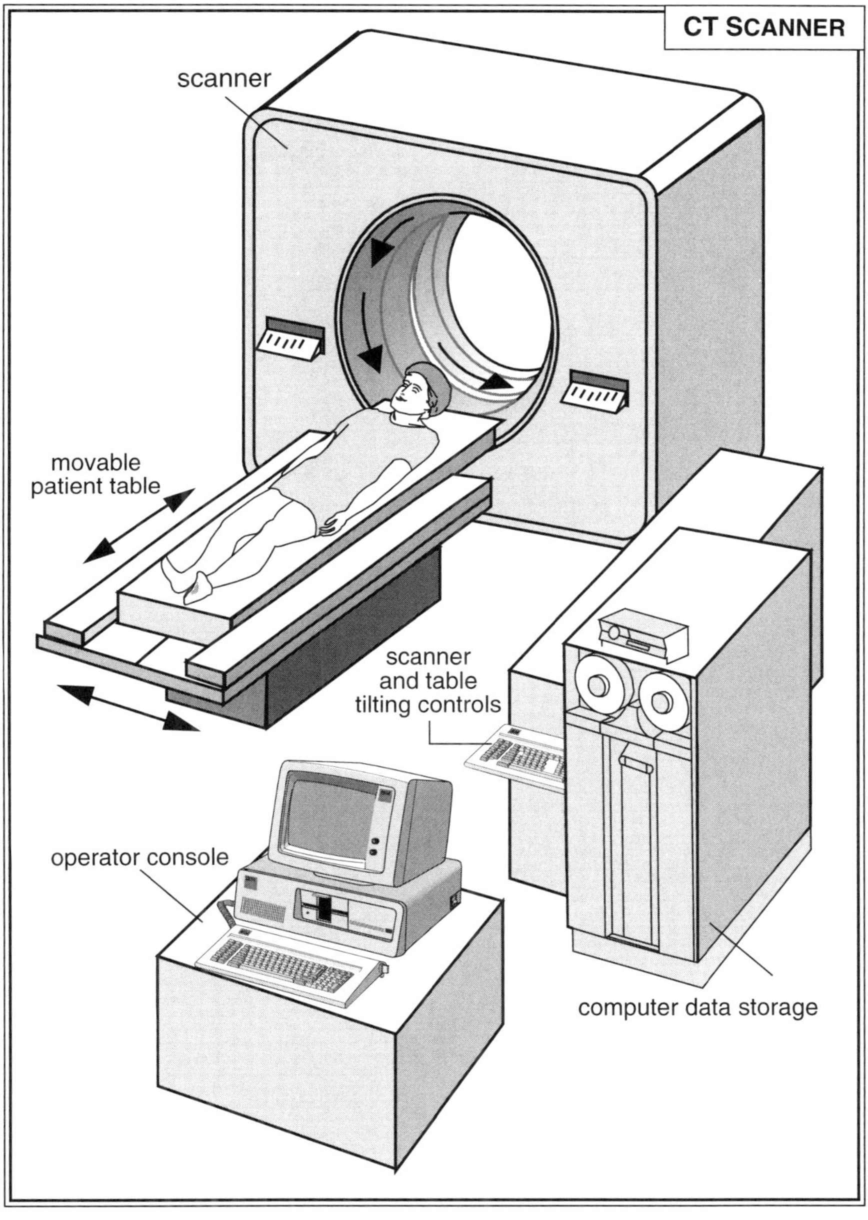

A patient having a CT scan lies on a movable table that can be passed through the hole in what looks like a gigantic square doughnut. X-ray sources in the inner ring of the doughnut produce a beam of rays that pass through the patient's body and are picked up by detectors on the opposite side of the ring. A computer turns signals from the detectors into an image that shows a cross-section or slice through the body.

through the body or are partly absorbed by the tissues. The ring containing the X-ray source and detectors can be rotated around the body, making thousands of readings. It can also be tilted to make a "slice" at any angle.

The results of the scanner's readings are digitized, or turned into a numerical form that the computer can read. The computer then combines them into an image, which the operator sees on a monitor screen. The scan takes 1 to 12 seconds and uses only as much radiation as a regular X-ray exam.

CT scanning has been further improved in recent years. For instance, regular CT scans show the beating heart or blood moving in blood vessels only as a blur. In the early 1990s, however, a new version of CT scanning was developed that is fast enough to freeze the heart in its tracks, just as very fast cameras freeze the motion of a flying bee or hummingbird. Ultrafast CT scans can show deposits of calcium in arteries around the heart, which are often an early warning sign of heart disease.

Atomic Tops

In the late 1970s, in addition to continuing to improve CT scanners, Godfrey Hounsfield began work on another new way of making images. It involved a technique called nuclear magnetic resonance (NMR), which chemists had used since the early 1950s to analyze the structure of complex chemical compounds. Today, NMR has joined CT as one of the "superstar" techniques of medical imaging.

In spite of the word *nuclear* in its name, NMR does not involve radiation or radioactivity. Instead, it is based on the way powerful magnetic fields affect the centers, or nuclei, of atoms. The nuclei of some kinds of atoms, such as hydrogen atoms, spin while tilted to one side, like tops. A strong magnetic field makes all the nuclear "tops" line up, just as a magnet makes iron filings line up. If the nuclei are then bombarded with a radio signal, they absorb energy from it and change the angle of their tilt. When the radio signal

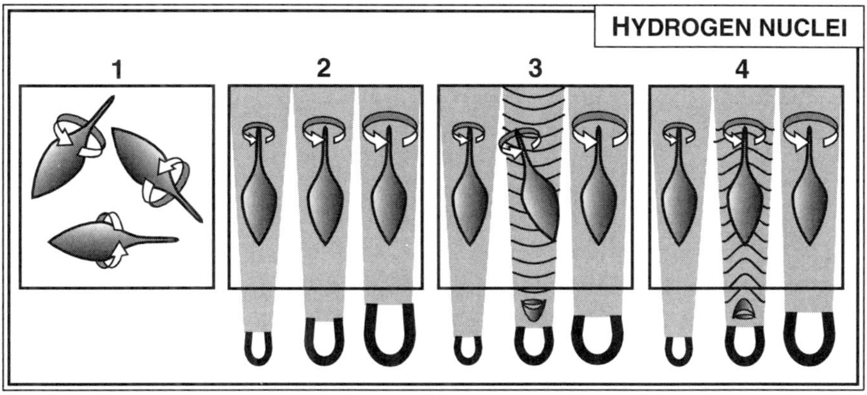

The centers, or nuclei, of hydrogen atoms spin like tops (1). If they are exposed to a strong magnetic field, they line up with the field (2). A brief radio signal knocks them out of alignment (3). When the radio signal is turned off, they line up with the magnetic field once again (4). In the process they give off a radio signal of their own, which can be detected and analyzed to show the location of these atoms in molecules.

is turned off, the atoms slowly go back to being lined up with the magnet. In the process of doing so, they give off their own radio signals. These signals show where the atoms are located in molecules.

Hounsfield suspected that, if he could make a device that could apply NMR to living tissue, it would help doctors understand chemical changes in the body. Different tissues ought to show different patterns of hydrogen atoms because the chemical compounds in them are different. For the same reason, cancers ought to look different from normal tissues.

Unknown to Hounsfield, an American doctor and inventor named Raymond Damadian had been working on the same idea since 1970. Like Hounsfield, Damadian thought NMR could be used to detect cancer. In 1970 he showed that NMR readings from rat tumors were different from those given by normal rat tissues. He then designed a NMR scanner that he thought could be used on living things. He won a patent on it in 1974. In 1976, he used an early model of the scanner to show a tumor in a live mouse. Most other scientists were not impressed, however. Some called Damadian's idea "visionary nonsense."

OTHER INVENTORS

CLAM SCAN

Scientists argue about whether Raymond Damadian or Paul Lauterbur of the State University of New York at Stony Brook first thought of the idea of using NMR to study living tissue. Damadian may have thought of the idea first, as suggested by the 1972 date on his patent application, but Lauterbur seems to have been the first to produce an image of a living thing. That first NMR "guinea pig," tested in 1973, was a 4-millimeter-long clam. The scan showed the clam's body inside its shell, and the clam survived the experience as happy as a—well—clam.

An Indomitable Machine

Damadian and his coworkers built the first NMR machine big enough to scan a human in May 1977. They nicknamed it Indomitable. No one was sure what effect a very strong magnetic field would have on a person's body, so none of the researchers was eager to lie inside Indomitable's central chamber and have his chest scanned. The group finally agreed that, as chief inventor, Damadian should have the debatable honor of going first.

Damadian took no chances. He had an emergency team standing by with equipment to restart his heart if the magnetic field should stop it, for instance. In fact, however, the scanning process did him no harm at all. Unfortunately, it also did not produce a usable image.

The explanation for the failure proved to be simple: The burly Damadian was just too fat for the size of the radio antenna coil wrapped around his chest. Luckily another member of the team, Larry Minkoff, was slimmer. It took six weeks, though, to persuade Minkoff to get into the machine.

The group tried their test again on July 2, 1977. This time they picked up a clear radio signal from Minkoff's chest. Minkoff then lay patiently in the machine for almost five hours while his body was moved forward, inch by inch. At the end of that time, the group had records of signals from 106 different scan points, arranged on a grid. Their computer then formed the information from these signals into an image, the first NMR scan of a human being.

In 1978, Damadian and his coworkers formed a small company called FONAR to make their scanning machines. They marketed the first MRI machines two years later. Several larger companies soon began making the machines as well.

Raymond Damadian was awarded the National Medal of Technology in 1988 for his invention. A year later, he was inducted into the National Inventors Hall of Fame. Still, he says, his biggest thrill was in "seeing [his] 'visionary nonsense' transformed into mainstream medicine."

From NMR to MRI

Because so many people associated the word *nuclear* with dangerous radiation, the name of the NMR process was changed to magnetic resonance imaging (MRI) during the 1980s. MRI machines were even more expensive than CT scanners at first, selling for a whopping $1.5 million apiece. Nonetheless, some 4,000 of them have now been installed worldwide.

Having an MRI scan is much like having a CT scan. Here, too, the patient lies on a table that is moved slowly into the center of the machine. An MRI scanner, however, is larger than a CT scanner. It eventually surrounds a person's whole body. An MRI scan usually takes longer than a CT scan—15 or 20 minutes. It is also more expensive.

One major advantage of MRI is that, unlike CT, it does not use radiation. It is therefore safer than CT. The strong magnetic field in the machine seems to do people no harm. They have to be sure to take off all metal objects before entering the scanning room, though. The machine's magnetic field, over 40,000 times as strong as the earth's, could snatch even a heavy wrench out of someone's hand. People have to leave their wallets outside, too. If they don't, the magnet will erase their credit cards.

MRI can form images of soft tissues and detect tumors even better than CT because, in an effect exactly opposite that of X rays, it does not show bones at all. It can also provide information about the chemistry of tissue that a CT scan cannot. It provides a very accurate way of studying the heart and brain.

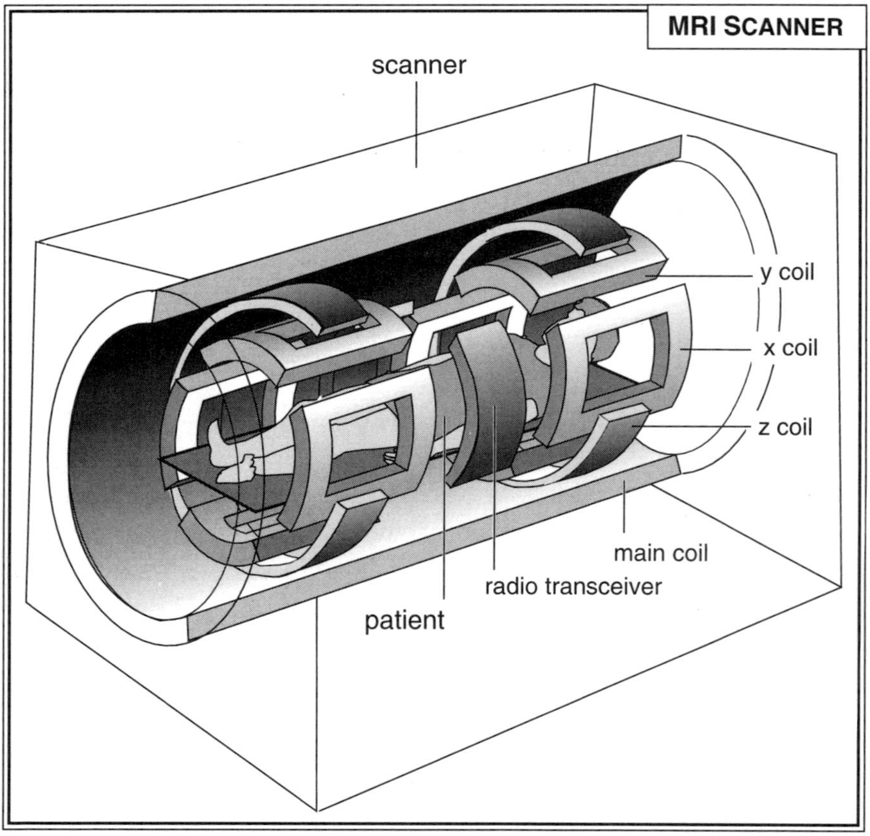

When an MRI scan is made, the patient's whole body is encased in the machine. The patient is exposed to several strong magnetic fields produced by magnetic coils, as well as to radio waves produced by a transceiver. The transceiver also picks up radio signals given off by atoms in the patient's body. A computer forms these signals into images that reveal much about the structure and chemistry of the patient's tissues.

I WAS THERE.

JOURNEY INTO AN MRI SCANNER

John J. Vollmer, a professor of chemistry, describes having an MRI scan made of his brain:

> With a push of a button, [a technician] moved the table and me into the heart of the [MRI] instrument. I felt like an astronaut wearing a special helmet and floating all alone in space. The helmet is equipped with mirrors to overcome feelings of claustrophobia [fear of enclosed spaces] and isolation. In one direction I could look down past my feet and out into the room. Looking straight up, I could stare into my own eyes, reflected just two inches away. . . .
>
> A strange staccato [sharp] hammering [began]. The . . . sound was very loud. . . . I compared it to a woodpecker's drumming, amplified hundreds of times. The noise was disconcerting but not awful, and I learned later that the sound originates from three additional magnetic fields that must be generated by electricity.

Digital Images

Other new imaging techniques are joining CT, MRI, and ultrasound as high-tech medical tools, and more are likely to follow in the next 10 or 20 years. Doctors will use them, as they do CT and the others now, not only to identify disease but to guide treatment. Medical imaging is used, for instance, to position beams of radiation aimed at tumors so that the greatest amount of radiation reaches the tumor and the smallest amount strikes normal cells.

TRENDS

COSTS OF MEDICAL IMAGES

The new medical imaging technologies are expensive, but as scanners are improved and more are installed, they are growing cheaper. The following figures show the costs of scans with the different technologies in 1993 and 1995.

Technology	1993 Cost	1995 Cost
ultrasound	$250–$500	$200–$400
CT	$500–$1,000	$350–$600
MRI	$1,000–$1,500	$500–$1,000

ISSUES

THE MAP IS NOT THE TERRITORY

Robert P. Crease, writing in the journal *Science*, warns that imaging technologies can cause problems if doctors do not use them carefully. One risk is that, because the images seem so realistic, doctors may forget that they are not the same as the body itself. Images can be inaccurate, just like any other test. A second danger is that the added information provided by the images may spur doctors into unnecessary action, such as performing surgery for a problem that is not really serious enough to justify an operation.

Increasingly powerful computers make imaging techniques ever more versatile. They can add color to emphasize certain features of a scan. They can blend many images to produce a three-dimensional picture. They can even combine images made with several different techniques.

As science writer Jenny Sutcliffe says, computers have even changed people's feelings about what an image is.

> Once, an image was a physical entity, a pattern of shadows on a piece of film. Now, it is more likely to exist as a pattern

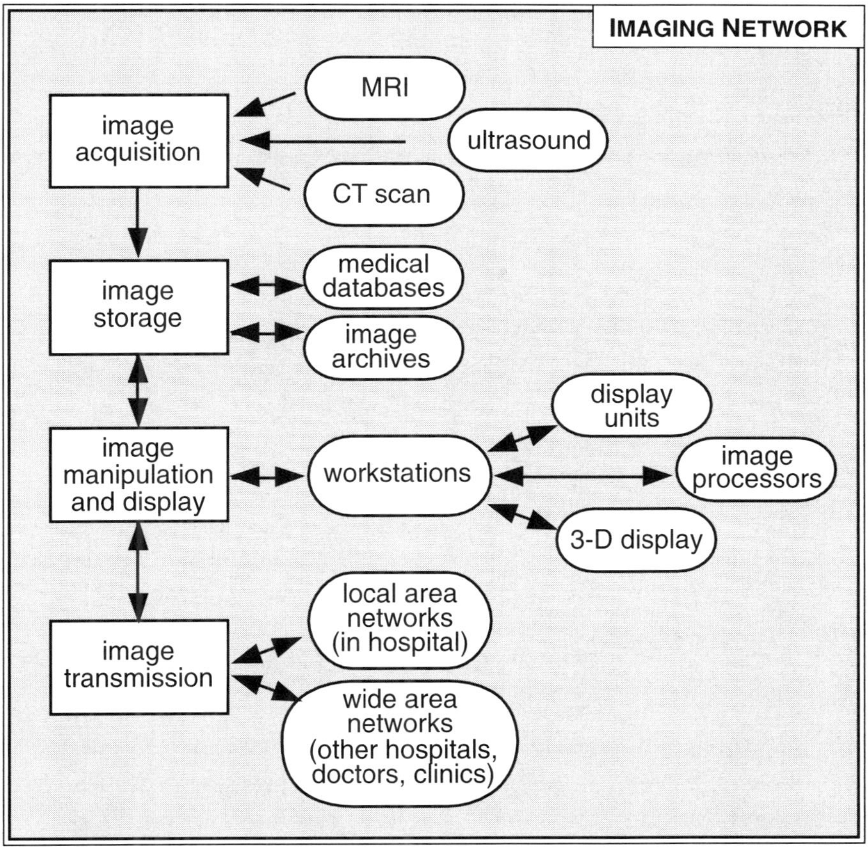

Most images used in medicine today are stored not as photographs but as information in a computer. This saves space, allows the images to be manipulated in many ways, and makes it easy to send them to distant locations.

SOCIAL IMPACT

THE COST OF "HIGH TECH"

Cost may limit the use of medical imaging. Because most new imaging tests are expensive, they help to drive up the costs of health care. Attempts to control such costs may force doctors to think carefully before ordering these tests in the future. "I think we've begun to [re]discover the less technological aspects of medicine," says Ronald Bayer of the Hastings Center, in Hastings-on-Hudson, New York, which examines ethical questions in medicine. "One of the reasons . . . is that we aren't able to afford all the high technology anymore."

of numbers held within a computer. Increasingly, imaging devices will be networked and their output electronically stored and distributed, potentially accessible anywhere in a country's health service.

Thanks to advances in medical imaging, the day when Dr. McCoy's hand scanner becomes a reality may not be so far off. Jeffrey Weinreb, professor of radiology at New York University Medical Center, says:

> Assuming cost does not become a limiting factor, you may end up seeing something not unlike what you saw on *Star Trek*. . . . Everyone will come into the hospital, or in for the annual checkup, and get their whole body studied and digitized and kept on permanent record, and you'll be able to look for problems in any part of the body. It . . . sounds futuristic, but the future is coming on faster than we think.

Chronology of Medical Imaging

August 28, 1919	Godfrey Hounsfield is born in Newark, England
early 1950s	Ian Donald begins work on applying ultrasound (sonar) to medicine NMR (MRI) begins to be used in chemistry
late 1950s	Ultrasound begins to be used in medicine
1967	Hounsfield gets idea for CT scanner
1970	Raymond Damadian begins attempts to apply NMR to living things
1972	First CT scan of a human being
1974	Damadian obtains patent for medical NMR scanner
1977	First NMR scan of human being
1979	Hounsfield receives Nobel Prize
1988	Damadian receives National Medal of Technology
early 1990s	Ultrafast CT scans developed

Further Reading

"Cosmo's Update on Medical Imaging." *Cosmopolitan*, September 1993. Overview of CT, MRI, and other common kinds of medical imaging, including description of the procedures, what they can detect, and their cost.

Damadian, Raymond V. "The Story of MRI." *Saturday Evening Post*, May–June 1994. Enjoyable description of the invention of medical MRI by the inventor.

Fincher, Jack. "New Machines May Soon Replace the Doctor's Black Bag." *Smithsonian*, January 1984. Description of CT, MRI, and other new kinds of medical imaging.

Hall, Stephen S. *Mapping the Next Millennium.* New York: Random House, 1992. Contains a chapter on use of medical imaging techniques to "map" the body.

"Hounsfield, Godfrey (Newbold)." *Current Biography 1980.* New York: H. W. Wilson, 1980. Interesting biographical sketch of Hounsfield, based on magazine interviews.

Sochurek, Howard. "Medicine's New Vision." *National Geographic*, January 1987. Well-illustrated description of medical imaging techniques, including what it feels like to undergo each one.

NOTES

p. 113 "I couldn't . . ." Quoted in Howard Sochurek, "Medicine's New Vision," *National Geographic*, January 1987, p. 40.

p. 117 "It's revolutionary . . ." Quoted in "Hounsfield, Godfrey (Newbold)," *Current Biography 1980* (New York: H. W. Wilson, 1980), p. 155.

p. 121 "visionary nonsense" Quoted in Kathleen McAuliffe, "The Inventor," *U.S. News and World Report*, January 26, 1987, p. 66.

p. 123 "seeing [his] . . ." Quoted in McAuliffe, p. 66.

p. 125 "With a push . . ." John J. Vollmer with Janet L. Hopson, "Looking at My Own Brain," *San Francisco Examiner/Chronicle*, April 10, 1988, "This World" magazine, p. 12.

pp. 127–28 "Once, an image . . ." Quoted in Jenny Sutcliffe and Nancy Duin, *A History of Medicine* (New York: Barnes & Noble, 1992), p. 232.

p. 128 "I think we've . . ." Quoted in Jack Fincher, "New Machines May Soon Replace the Doctor's Black Bag," *Smithsonian*, January 1984, p. 68.

p. 128 "Assuming cost . . ." Quoted in Stephen S. Hall, *Mapping the Next Millennium* (New York: Random House, 1992), p. 154.

Epilogue

Medical Technology in the Twenty-first Century

Medical technology has made amazing advances in the 19th and 20th centuries. Even more amazing developments are likely to occur in the century to come.

Like those in the late 20th century, many of the 21st century's advances in medical technology will arise from the ever-increasing power, speed, and miniaturization of computers. "You can think of medicine as just another aspect of the information age," says army surgeon Rick Satava. "What we used to think of as blood and guts are just bits and bytes. . . . It's transferring information back and forth."

Satava says that the 21st century will be the age of the "digital physician." As he explains, "All medical information . . . can . . . be brought to the physician or surgeon in digital format. Now we just close the loop and let the surgeon manipulate and act on the patient through the digital world of information."

Surgical Video Games

Some types of surgery are already being turned into a kind of video game. Surgeons guide their hands by what they see on a monitor screen rather than by what they observe directly in the patient. Some of these images come from real-time CT, MRI, or ultrasound scans. Position sensors on the surgeon's instruments allow the instruments as well as the patient's tissue to show up on the scans.

Other images come from computer chips and tiny cameras embedded in devices called endoscopes. These can be inserted into the body at the end of long tubes slowly pushed through body openings or through small surgical incisions. The cameras send pictures back to the surgeon by means of light-carrying fibers (fiber optics).

Using fiber-optic endoscopes, surgeons can do some kinds of abdominal operations through tiny incisions in the navel, rather than having to make wide incisions that leave a scar. Patients recovering from such operations often leave the hospital the same day and return to work a few days afterward. Before modern endoscopes were used, people had to remain in hospital beds for several painful weeks while their incisions healed.

In the future, an operating room may look even more like a video arcade. Before making an incision, a brain surgeon may point a wand at the spot on a patient's head where he or she plans to begin cutting. The wand is connected to a computer. The computer's monitor screen will then show the inside of the patient's head (based on earlier imaging scans) from that angle. The surgeon will be able to "practice" the operation in three dimensions, moving the image at will, before performing it in real life. This will allow the surgeon to plan the operation very precisely, reducing the size of the incision and the risk of damage to healthy parts of the brain. Wand devices of this type were first developed in the early 1990s.

Robot Surgeons

Some surgeons already let robots do part of their work. The surgeons program the robots' motions by computer before the operations begin. In 1985, for instance, Yik Sin Kwoh of Long Beach, California, used a robot to take a sample of suspected tumor tissue from a woman's brain. "The robotic arm is safer, faster and far less invasive than current surgical procedures," Kwoh said. A few years later, William Bargar of Sacramento, California, used a different robot, which he calls Robodoc, to drill a hole in a patient's thighbone. An artificial hip joint was then placed in the hole. Such a hole must be shaped very precisely to fit the new joint. Bargar claims that "Robodoc"'s drilling is 20 times more accurate than that of a human surgeon.

In the future, thanks to computers and robots, surgeons may not have to be in the same room or even the same hospital as their patients. A surgeon will grasp the handles of instruments attached to a console and carry out an operation on a three-dimensional image on a computer screen. The surgeon's console will be linked by a computer network to the operating room where the real patient lies. Robot "slave" devices in the distant operating room will repeat the surgeon's movements on the living patient. "Telesurgery" like this could let a top surgeon operate on patients around the country without either surgeon or patients having to travel. It could also help surgeons operate quickly on people injured in battles or natural disasters.

Computers will help surgeons overcome obstacles of size as well as distance. Just as surgical microscopes now magnify a surgeon's vision, computerized robot devices in the future will let the surgeon "shrink" his or her hands. When the surgeon makes a 1-inch (2.54-cm) motion, for instance, a "slave" robot may move as little as a billionth of an inch. Some devices planned for the next century have microrobots that can operate on a single cell.

Computer Help for the Disabled

Computers will also play key roles in future devices to help the disabled. They already are an important part of such devices as

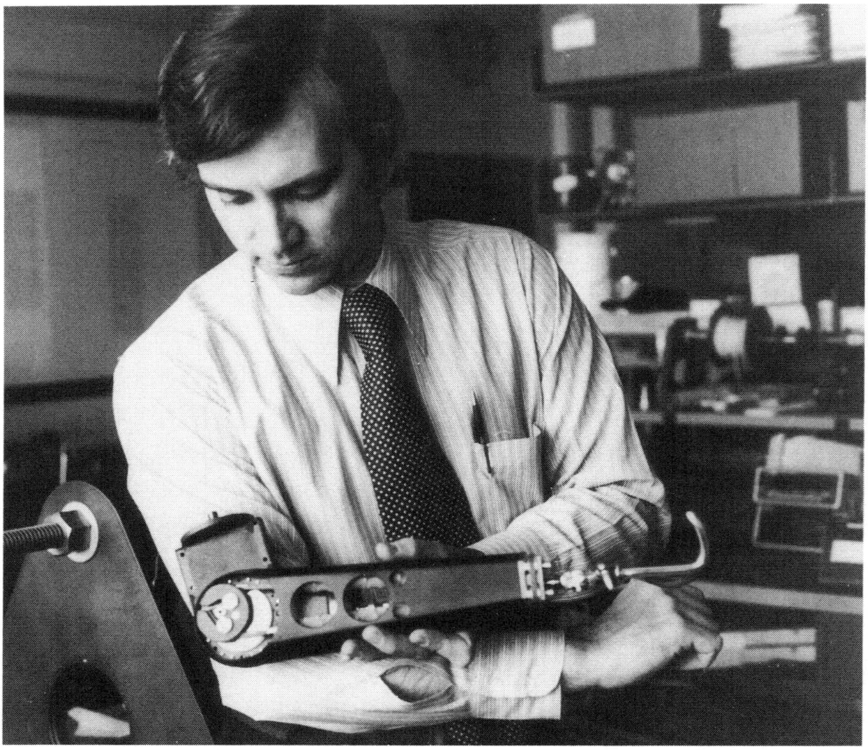

Stephen Jacobsen designed the Utah artificial arm in Willem Kolff's laboratory at the University of Utah. The arm uses a computer chip in the elbow to translate signals from the wearer's muscles into actions performed by electric motors in the artificial arm. People wearing the arm can carry out a wide variety of actions. (Courtesy Manuscripts Division, University of Utah Libraries)

the Utah arm, developed by Stephen Jacobsen in 1981 in Willem Kolff's department at the University of Utah. Tiny electrical signals from muscles in the amputee's stump control the arm's electric motors with the help of a microchip in the elbow.

Alice Olson, the first person to wear a Utah arm (combined with an equally advanced device called the Otto Bock hand), said, "When they fitted me, I got tears in my eyes. The arm's response was so natural, it was almost like having my own arm back." A reporter says that Olson can use her Utah arm and hand to "hold eggs or Styrofoam cups of coffee, open doors, carry luggage, push carts, . . . crack walnuts, slice tomatoes, peel vegetables, open jars, knit, ride a bicycle, and put on makeup."

In the future, computer chips linked to nerves will allow an even greater range of activity and sensation. One new kind of artificial hand, for instance, gives its wearer a sense of touch. "The first time I could reach out and touch my wife's hand and feel the warmth after more than a decade—that was a very emotional moment," says Chuck Tiemann, who got one of the new hands in 1995, 15 years after losing his arm in an accident.

Computers are even being taught to read brain waves so that completely paralyzed people can use them. Electrodes attached to the scalp pick up the brain waves and send them to a computer. The computer translates these signals into instructions for moving a cursor on a monitor screen. Depending on the way the computer is programmed, the movement will produce communication or signal devices to perform simple tasks, such as changing the channel on a TV set.

Engineering Human Genes

Advances in the field of genetics will lead to other new developments in medical technology. Scientists have already discovered genes that cause many kinds of inherited illness or increase the risk of developing diseases such as cancer. The Human Genome Project, scheduled to be completed in 2003, intends to map the entire collection of genes (genome) in human beings. Leroy Hood, inventor of a device that automatically analyzes genes, says that once the functions of all these genes are known,

> when a baby is born, we'll "read out" his genetic code, and there'll be a book of things he'll have to watch for. This has the potential to do enormous good. If you have a propensity [tendency] toward heart diseases . . . or cancer, you could modify your diet or change the environmental substances you're exposed to [to decrease your risk of getting the disease].

Once genetic testing technology identifies a damaged gene, the technology of gene therapy may be able to replace that gene with a healthy one. Still in its infancy, gene therapy so far has been used to correct only a few rare conditions caused by a defect in a single gene. In the future, it may be used to treat complex diseases such as cancer, heart disease, and AIDS.

Genetic engineering and the related field of biotechnology are also likely to help medicine in more indirect ways. For example, scientists have used genetic engineering to make bacteria, mice, and even cows produce human hormones and other substances that are scarce in nature. Other genetic engineers have put human genes into animals so that the animals can be used to test treatments for human diseases or provide possible sources of organ transplants.

Challenges to Medical Technology

Even as it advances, medical technology will face severe challenges in the century to come. Chief among these is likely to be cost. Health care costs take up an ever-growing percentage of the Gross National Product in the United States and other developed countries. These costs strain economic systems and use up money that might be spent to meet other human needs. The cost problem is likely to become even worse in the future because people in developed countries are living longer and longer. Old people often need more health care than young people.

Private insurers, the government, and individual patients are all eager to control health care costs. One way to cut costs is to reduce the use of "high-tech" medicine, which is often expensive. For instance, people may decide that health care dollars should be spent to provide preventive care to large numbers of relatively healthy people rather than to give artificial organs or transplants to a few severely ill people. Whatever choices are made, the decisions are sure to be both hard to make and controversial.

Another challenge comes from people who question the whole way of thinking that focuses on technology. The technological approach, so large a part of medicine in the 20th century, not only uses machines to examine or repair the human body but tends to see the body itself as a sort of machine. It tries to repair or replace diseased body parts much as a mechanic might do with broken machine parts. Such an attitude, its critics say, keeps doctors from seeing patients as whole individuals. These critics point out that health or disease usually arises not just from single body parts but from all the interacting systems of the body, including the mind. Many illnesses may be better treated by changes in attitude, environment, or lifestyle than by drugs or surgery.

Medical technology will surely go on being developed and used in the next century. Human health will be much the poorer if it is not. Doctors may learn to use it less often, however, and consider less expensive and less mechanical approaches instead. Rather than being the chief tool of medicine, medical technology may become just one option in an integrated approach that weighs costs, benefits, and an understanding of the individual in an attempt to find the most effective medical care for each person a doctor sees.

Further Reading

How Things Work: Medicine. Alexandria, Va.: Time-Life Books, 1991. Well-illustrated book describes current and near-future medical technology, including advances in surgery, computer diagnosis, and devices to help the disabled.

SRI Center for Medical Technology: Telepresence Surgery. http://os.sri.com/medical/telepres.html. Has overview of research in telepresence (remote-control) surgery.

Taubes, Gary. "Surgery in Cyberspace." *Discover*, December 1994. Describes predicted advances in surgery, including possibility of performing surgery on patients in distant locations using "virtual reality" techniques.

Yount, Lisa. *Genetics and Genetic Engineering*. New York: Facts On File, 1997. For young adults. Describes recent advances in these fields, including those affecting medicine.

NOTES

p. 131 "You can think . . ." Quoted in Gary Taubes, "Surgery in Cyberspace," *Discover*, December 1994, p. 85.

p. 132 "All medical . . ." Quoted in Taubes, p. 86.

p. 133 "The robotic . . ." Quoted in "Robot 'Surgeon' Operates on the Brain," *San Francisco Chronicle*, June 25, 1985.

p. 135 "When they . . ." Quoted in Janice M. Cauwels, *The Body Shop* (St. Louis: C. V. Mosby, 1986), p. 64.

p. 135 "hold eggs . . ." Cauwels, p. 64.

p. 135 "The first time . . ." Quoted in "Amputee Feels Renewed Hope with 'Bionic' Limbs," *San Francisco Chronicle*, April 28, 1995.

p. 136 "when a baby . . ." Quoted in Joel Davis, "Leroy Hood: Automated Genetic Profiles," *Omni*, November 1987, p. 118.

Index

Page numbers in *italics* indicate illustrations. Page numbers in **boldface** indicate major treatment of topics.

A

L

M

N